APERTURE

Exploring what the central mystery of the bond between women and across generations might be, *Mothers & Daughters* marks the first photographic delineation of the emotionally laden, socially explicit relationship between mothers and daughters. The images presented here are ample evidence that, as mothers or as daughters, women are not alone: in passion, conflict, confrontation, and reconciliation; and in the diversity of our life-styles and racial and ethnic identities.

These 84 photographs are part of an exhibition of 130 photographs touring the United States over the next several years that explores how women live and act, chronicling their most intimate expressions to each other: about ambition and achievement, conscience and politics, family, marriage, and relationships, sex and physicality. In its small details and overall vision, this collection of pictures charts new ground, illuminating those private moments of women's lives with the intimacy of a diary and the authority of many voices.

In making the selection from nearly three thousand photographs, the editors considered the similarity in values, and the difference in perspectives, work, and decisions represented by the many images. In the balance, a richer and more eloquent picture emerged than anyone could have expected. There are many familiar names among the photographs included, as well as significant new talent.

Tillie Olsen provides a deeply thoughtful memoir, coauthored by her daughter Julie Olsen Edwards. The Olsen and Edwards text provides a counterpoint to Estelle Jussim's provocative examination of the visual history and conventions of mother-daughter depictions. *Mothers & Daughters* also includes a selection of texts from a variety of well-known women writers and poets: Sharon Olds, Marge Piercy, Denise Levertov, Nikki Giovanni, Alice Walker, Maxine Kumin, Adrienne Rich, Margaret Mead, Eudora Welty, Kathleen Spivack, Susan Minot, and Anne Sexton, as well as observations from a number of the photographers.

Historically, the mother-daughter relationship has been either ignored or trivialized. It is a story sometimes obscured by sentimentality, or seen only through the externalized and internalized pressures of culture. It is silent, this history: "We know nothing of them except their names, the dates of their marriages and the number of children they bore," as Virginia Woolf wrote.

Resonant with many faces, many voices, *Mothers & Daughters* is an anthology of conflicting, idiosyncratic, individualized visions, mirroring a quality of connectedness, of warmth as well as pain. Transmitted and documented here, across generations, from mother to daughter and daughter to mother, is the desire for fulfillment, the longing to preserve security, and the ferocity of the need for independence.

THE EDITORS

Work by the following photographers, in the order in which they appear, is represented in this issue:

Bea Nettles, Kathleen Kenyon, Bruce Davidson, Niki Berg, Starr Ockenga, Maude Clay, Rosalind Solomon, Joel Meyerowitz, Keith Glasgow, Stephen Scheer, Eudora Welty, Laura McPhee, Roland Freeman, Tom Bamberger, Leon Borenzstein, Joyce Baronio, Sally Mann, Mary Kalergis, Linda Brooks, Judith Black, Susan Jahoda, Mark Berghash, Barbara Crane, Mark Goodman, Sage Sohier, Bruce Horowitz, Arthur Tress, Frances McLaughlin-Gill, Larry Fink, Charles Harbutt, Hope Wurmfield, Raisa Fastman, Carla Weber, Abigail Heyman, Randy Matusow, Nick Nixon, Jason Laure, Jill Freedman, Danny Lyon, Paul Fusco, Michal Heron, Herman Leroy Emmet, Carl Richter, Joan Lifton, Elaine O'Neil, Joseph Szabo, Susan Copen-Oken, Elliott Erwitt, Raisa Fastman, Bill Owens, Arlene Gottfried, Margaret Randall, Aneta Sperber, Rosemary Porter, Milton Rogovin, David Graham, Todd Merrill, Todd Weinstein, Eric Breitenbach, Jock Sturges, Rhondal McKinney, Robert Adams, Harry Callahan, Garry Winogrand and also: Matthew Brady, Henry Peach Robinson, Gertrude Käsebier, Lewis Hine, James Van der Zee, Michael Disfarmer, Dorothea Lange, Eve Arnold and Jerome Leibling.

Kathleen Kenyon, *Mother Teacher (series)*, Woodstock, New York, 1984

MOTHERS & DAUGHTERS

THAT SPECIAL QUALITY

AN EXPLORATION IN PHOTOGRAPHS

ESSAYS BY
TILLIE OLSEN WITH JULIE OLSEN EDWARDS
ESTELLE JUSSIM

AN APERTURE BOOK

What is astonishing, what can give us enormous hope and belief in
a future in which the lives of women and children shall be
amended and rewoven by women's hands, is all that we have managed
to salvage, of ourselves, for our children . . . the tenderness, the
passion, the trust in our instincts, the evocation of a courage
we did not know we owned, the detailed apprehension of another
human existence, the full realization of the cost and precariousness
of life. The mother's battle for her child—with sickness, with
poverty, with war, with all the forces of exploitation
and callousness that cheapen human life—needs to become a common
human battle, waged in love and in the passion for survival.

ADRIENNE RICH

5 Bruce Davidson, *Women with Child in Subway*, New York, New York, 1979

 Niki Berg, *Joan and Julie*, Hancock, Massachusetts, 1981

7 Starr Ockenga, *Verena and Francesca*, Boston, Massachusetts, 1981

Brushing out my daughter's dark
silken hair before the mirror
I see the grey gleaming on my head,
the silver-haired servant behind her. Why is it
just as we begin to go
they begin to arrive, the fold in my neck
clarifying as the fine bones of her
hips sharpen? As my skin shows
its dry pitting, she opens like a small
pale flower on the tip of a cactus;
as my last chances to bear a child
are falling through my body, the duds among them,
her full purse of eggs, round and
firm as hard-boiled yolks, is about
to snap its clasp. I brush her tangled
fragrant hair at bedtime. It's an old
story—the oldest we have on our planet—
the story of replacement.

SHARON OLDS

Andra was thirteen then, the only daughter in the family. It was July in Memphis, late afternoon of a hot, sticky day, and they had been fixing up the house when I drove up. Maybe something happened before I got there, maybe it was just a moment when you feel like hugging someone you know. MAUDE SCHUYLER CLAY

 Maude Schuyler Clay, *Rosa and Andra Eggleston*, Memphis, Tennessee, 1984

It was summer, in 1984. I was wandering on the sand at Brighton Beach that day, and they just looked so wonderful that I stopped and asked if I could take their picture; then they turned to one another like this—it just happened. . . . It is some special kind of nonverbal connection. I always hope for that in my pictures, something intense, that happens for an instant. . . . ROSALIND SOLOMON

 Rosalind Solomon, *Untitled*, Brighton Beach, New York, 1984

11 Joel Meyerowitz, *Stella and Tessa*, Provincetown, Massachusetts, 1985

 Keith Glasgow, *Soldiers*, Fort Dix, New Jersey, 1985

 Stephen Scheer, *Jacob Riis Park*, Queens, New York, 1984

MOTHERS & DAUGHTERS

Tillie Olsen with Julie Olsen Edwards

Mother, I do not know you. Mother, I never knew you. Daughter—you knew yourself. Without knowing, you knew me.

Here, for the first time in all photography's history of book collections and museum exhibits radiating around a theme, the subject is Mothers and Daughters—that crucial relationship still veiled in the unseen, the unexpressed, the unarticulated; still swathed by societal precept, presumptions, pronouncements.

Such an assemblage—way making, pioneer—was not intended to be, could not be expected to be, a coherent essay exploring this tormentingly complex relationship. Inevitably, there is an individual range of interest, motivation, focus. There are primary aesthetic, pictorial values; considerations.

But one hundred thirty images; nearly all of them U.S.A., contemporary . . .

The eye—enabled—seeks vision.

Here are daughters and mothers of every shape and human hue; in every age and stage from mother and infant, to old daughter and old, old mother; and here is the family resemblance in face, expression, stance, body.

Here are mothers and daughters of lack and of privilege, in various dress, settings, environments; posing for photographs or (unconcerned with the camera) sharing tasks, ease, occasions, activities; holding, embracing, touching; or in terrible isolation.

Here is sullenness, anger or controlled anger, resentment; admiration, distaste; playfulness, pride; joy, joy, joy in each other; estrangement; wordless closeness or intense communion.

A welter of images. Multi, multi-form.

The eye seeks deeper vision.

What, besides appearance (the obvious femaleness of dress, and, after a certain age, of body) distinguishes this as pertaining especially to mothers and daughters? (Family resemblances; the changes and weathers of time; the visual demonstration that human relationships inhabit various settings, life circumstances; the emotional range—these would also be true of a mothers and sons, or daughters-fathers, fathers and sons, collection.)

What, specific only to mothers and daughters, is profoundly, pervasively present—but encoded, hidden here?

—Why the prevalence of portraits, of "stills"—charged as they may be with quality of mood or relationship, socio-economic class; visually beautiful as many of them are in the traditional ways mothers and daughters are pictured together?

—Why are there comparatively few "action" photographs, little of the everyday acts, the experiences which create, condition the relationship?

—Why do certain images or juxtapositions seize one with memory, recognition; a troubling sense of contradiction or lack; a trembling on the edge of comprehension?

The eye seeks vision.

What leaps out at once from frame to frame, singly or chorded together, is the overwhelming number of photographs where there is holding, touch, embrace; bodies having to do with each other, a tangible sense of connection.

Even when physically apart, or distanced by mood, other involvement, almost always there is that bodily sense of connection. Even—notably—in the images of abandonment, estrangement, hostility, their very intensity reveals how deep the mattering; the seeming unattainability of indifference; the indissolubility of some kind of bond (tie, yoke).

Out of the beginning years? (The eye notes the manifold maternity/child (Mary Cassatt) images of mother adoration, delight, pride; child sensuous bliss, contentment.)

Bodies once so known to each other:

—By the mother, in the tending, the caring, so necessarily physical in nature; in the very feel, downy softness, weight, scent of the child's body; in her response to its blossoming needs;

—By the child, as warmth, soft rotundities; needs met, pleasured by the mother's hands, face, voice response; ineradicably experienced in the ecstatic unfolding of infant senses and powers, in the very shaping of memory.

Secret deposit of happiness, of safety, comfort, assurance; of fostered blossoming.

Child/Mother. But only between mothers and daughters (not sons) is it legitimized: that lifelong at-homeness with each other's bodies, the sensuality, the easy tactile expression of connection.

Only the *possibility*. What is not legitimized, enabled, is the continuance *beyond* the early years of what the bodily closeness and caring, at-homeness, once seeded and now symbolizes: mutual need and fostering; understanding and joy in each other; abiding connection.

(Indeed for the grown daughter, such attachment is defined as immaturity, dependence; the inability to achieve autonomy. For the mother, it is seen as "holding on", interfering, "being too close.")

That in reality, (and it is implicit in photograph after photograph here) daughter and mother might be, often *in spite of all* are: agents of mutual growth, encouragers of each other's independence and capacities, lifelong comrades on the long journey, this reality and its possibility is obscured, impeded, denied.

In spite of all. Evasive, incoherent "all": that thwarts, hinders, comes between; (mis)interprets, poisons.

The eye seeks out the "I rejoice in you, you rejoice in me" mother-daughter embraces at every age, along with the contrasting images of estrangement, of hostility, of bitterness. Disparate, connected. Not yet fully amassed into comprehension.

Some necessary generalizations towards comprehension: (in an inescapably sociological vein):

—Motherhood is idealized, mythified, sentimentalized, yet (twentieth-century phenomenon), it is ridiculed, indicted, blamed.
—Mothers are expected to be the center of love and health the outside world is not—but seldom, if ever, are they empowered.
—The actual work, content, worth of mothering (or any other "women's work") is seldom visible—whether in the arts (including this photographic exploration), literature, or society's considerations.
—It must usually be fit in with other responsibilities, cares, needs—including our own fulfillment-needs.
—It is unpaid—in moneys. Its profound joys, fulfillments, are suspect, if they are evidenced at all.
—Only a fraction of possible contributions to the other fields of human thought and endeavor are ever made—not because of having children, but because of the circumstances in which children must be raised.

The task of mothering daughters is different than that of being mother to sons:

Regardless of their illimitable endowment as born human, the little ones must be shaped (mis-shaped) into being female or male, as society prescribes it.

Whether we assent, attempt to ignore, resist or transcend it, there is still a reality of "man's world, woman's place." Integral to that "woman's place" are damaging economic and other inequities; almost sole responsibility for the most necessary, demanding (yet devalued) maintenance-of-life tasks; mandated behavior for girls, for mothers, for women; . . .

The arena in which, from the beginning, these realities are felt, acted out, exercised, resisted or surrendered to, is the mother-daughter relationship. It is permeated throughout with these special circumstances.

The strain, stain, weight, toll is enormous. Few escape it.

Blessed Art Thou Among Women

Beautifully adorned for entrance/exit, contrasted by the beneficent curve of her mother beside her, the daughter in the Käsebier portrait (page 100) stands in an open doorway, in a dazzle of light, of possibility . . . .

Her just-budding body curved into the swollen bulk of her pregnant mother's belly, her arms flung around her mother in clinging, passionate protectiveness, Sally Mann's Jenny (page 25) looks out at us unforgettably: in wistfulness, in fear; in a compound of too-premature understandings, and wants, and cares . . . .

Arrayed in the homemade Bicentennial Celebration dress like the one her grandmother has made for herself (page 20), her grandmother's arm proudly around her, Tammie Pruitt Morgan asks questions of us—radiates solidity, intelligence, shared strengths . . . .

Basking in the center of admiration and attention, the curve of her mother's hand arranging her hair like a benediction (page 44), the blissful girl-child awaits her First Communion in a web of dreams, imaginings, aspirations . . . .

And the occasional apprentice-colleague daughters herein, sharing work, learning skills; and Maude Clay's Andra Eggleston and Rosalind Solomon's unnamed girl-child (arms encircled, encircling) and Jock Sturges' Sandy and Catherine, Starr Ockenga's Verena (beached against the bodies of their mother-beings who have instilled into them what should be everyone's legacy: the sturdy personhood out of having been cherished, joyed in, encouraged to burgeon (pages 87, 9, and 10, and 93) . . . .

And each of the other (not always so fortunate) named, or nameless, girl-child daughters becoming women that we view here . . . .

World, be tender with her . . . World, don't dim, blight,
destroy . . . Let her flower in fullness of worth, of being

Judgmental class, race, world. World of economic imperatives;
of rape, of violence; of premium on appearance, on young
sexuality. "Man's world." World not concerned with
human worth, human flowering.

She will hold it against me, perhaps always
That I could not magic it to be right for her
That I was not able to prepare her, to arm her

Rage and betrayal and hurt. *Against me.*

It is my voice (and being) that must try to translate society's standards, realities, imperatives, to her—and it is I who will receive her despair, anger, rebellion over those lessons, over those realities.

Worse: mine may be the situation in which (having no choice) like the mothers with bound feet in old China, who were required to bind the feet of their daughters; knowingly or unknowingly, willingly or unwillingly; I must bear complicity in fitting her into those maiming realities. . . .

Those terrible realities, at war with her needs, her capacities, her potentialities.

The war in her;
The war in me, too.
Ally, or foe, or both?

The tender pride in her young worth, her promise—like our old tactile pleasure in washing, brushing, braiding her hair; seeing her "presentable"—have a corrosion breeding in them now, that of the world's standards: too thin, too fat, too tall, too short, too smart, too dreamy, too athletic, too serious, too this, too that;

It is the time when obsession with appearance, body, clothes begins (compelled obsession for females—but she may not know that yet, if ever). The blossoming being must be gendered.

They have eleven, twelve, thirteen, fourteen-year-old bodies; pregnable women's bodies in a country where sexuality is accorded primacy—where the hunger for belonging, independence, for verification of selfhood, the need to test oneself in the unknown, is too often channeled into the sexual arena.

How to get through these years, like and unlike the years we once blundered through (and gloried in); how to help when needed—without harming; how to fight for her to develop in the fullness of her vital powers;

How to be ally—not foe. How not to lose her.

It is these years when estrangement becomes common. *"Out of her womb of pain, my mother spat me, into her harness of ill-fitting despair . . . her nightmare of who I was not becoming".*[1] The photographs of daughter—sometimes mother—resentment, sullenness, coldness proliferate.

Estrangement around constraints (if we try to place them); around our hopes and expectations; my mother's *"concern about my life, which I needed and took for granted, {but} could not bear to have expressed;"*[2] estrangement around her unmet needs—not alone that we have to be at other tasks, cares, needs—but out of her encounters with that *world not concerned with human worth, human flowering. . . .*

Wait. Whatever happens between us there is another truth; a solace: my secret deposit. Out of the beginning years (and often after) of caring, being cared for, cared about; ineradicably deep in most of us, is a deposit of strength: inexhaustible source of resilience, courage, tenacity, hope—of belief in inalienable human right (capacity) for life, liberty and happiness.

(But she may never come to consciously credit, know, this.) (Nor I.) (Nor society.)

Again, the eye dwells on images of grown daughter/mother estrangement, hostility, sense of hungering needs unmet or unperceived . . . disparate/connected; ponders the photographs suggestive of the sources of mother/grown daughter abiding closeness (*Unbroken Bonds*), so consummately conveyed in Niki Berg's *Self-portrait with Mother* . . . Those colleagues in work, the singing Judds, *Naomi and Winnona in Concert*, the body-building Paynes; the instances of mutual endeavor, shared beliefs, joyous activities: Raisa Fastman's "Wonder Women", gussied up in all their regalia, Patricia Evans' white-haired mother with her cherishing hand on her daughter's pregnant belly, Margaret Randall's bikeriders; the bannered mother and daughter at the Washington ERA demonstration, more subtly, the intensity of the listening, the communication between Larry Fink's Henrika Frajlich and her mother, signs of deepest mutual involvement; encoded intimations, clues.

[1] Audre Lorde, from "Story Books on a Kitchen Table," *Coal*, W. W. Norton, 1976
[2] Alice Munro, from *Lives of Girls and Women*, McGraw-Hill, 1971

Daughters/mothers. Estranged or close, in the grown years, there are similarities between them now: women's bodies, no matter how differently we may inhabit them; women's lives, no matter how differently they may be led;

Whatever other work she does in this world, whatever other choices she makes (if she has choices), whether mothers or not—almost inevitably a daughter will end up, as her mother did, as female: maintaining family and personal connections; cleaning, cooking, providing primary health and other necessary care; creating holidays, celebrations of birthdays and other milestones, feeling joy at seeing happiness on the faces of those we love.

Replicating her mother's life in other ways. . . .

Sometimes becoming mothers, too (and having that in common—both were daughters, are mothers). . . . And the shift, the change in what we understand of our mother's life as we begin to (or try not to) repeat her experience. Whether we know our mothers with joy or with bitterness, all of us rethink, in a way relive, her life as we mother . . .

Realizing the hours of caring, of mediating, of making our world more safe, manageable, joyous; the sheer physical work of mothering: carrying, lifting, balancing a child on one hip, groceries on the other, running after the little ones; the sleeplessness; the myriad of details to attend to; the piece of the brain that no matter what else eclipses it, is always concerned with the well-being of the children (what is happening at school, are they alright, what is there for dinner).

Whether we mother or not, we do not, can not as adults, remember the details, hours, tasks of our mother's work, only isolated moments out of the years of care. Nor can we (how could we?) remember her as the person—besides our mother—she was in those years.

Then there comes a time when, with most of us (*all that has been before: the deposit; our own experiencing*), we begin to consciously understand the truth of our mother's life—how hard she has worked, her struggle against limitations, what life has done to her body; begin to comprehend her as a human being in her own right.

She is old. On her face we see the process of aging (the portent of our own aging), the price of the way of her living, the awareness of what was not possible, did not happen. We can see the strength of that which was good and that which was lived over, survived, even transcended, transmuted.

Perhaps we recognize that in her, too, is the continuing need to grow, experience, live, as deeply, fully as possible; recognize that what mauled, limited, perhaps distorted her may be limiting, mauling, denying us.

We may both share the realm of the unresolved, the unfulfilled, the unattained—and the healing recognition of the compass, the beauty, of what we did attain.

(It is the time when photographs of our mother when young haunt us; pierce us with undefinable emotion.)

I use to be a Mother to my daughters. Now I like a baby to them . . . (from Jim Goldberg's series, immeasurably constellated by the visual presence, in their own handwriting, of the photographed) . . . *I hate to see my Mother like this . . . I wish I could have kept her home with me like a Rose Kennedy . . .*

—The time of the older mother's need—need we are rarely able to fill. The terrible burden of knowledge if her life is not good. And sometimes she has no one else. The tenderness, love for her we may never have realized was in us so often warring with our wanting to be free—while we are in her old maelstrom of mid-life work, responsible for other dependent ones.

—Her own hatred of having to be dependent deepened by this knowledge . . .

Daughter, Mother *Mother, Daughter*

This assemblage, way-making, pioneer—with its treasures, beauties, sometimes irrelevances—is our first. Vista upon vista opens. So many facets of mother-daughter relationships are yet to be imaged. So much—directly out of our time—is yet to be suggested, situated, illuminated.

O fortunate photographers! O challenged photographers! How to make more visible for us "the everyday acts, the experiences which create, condition the relationships"; how to indicate the societal circumstances in which the mother-daughter relationship must exist, how to help us see the sources of estrangements, the roots and ways of abiding closeness.

Mary Frey with her *Real Life* series; Rosemary Porter in her documentation of her working self, of her daughter (complete with her daughter's eloquent notes), Sally Mann, and other photographers here are portent of what mothers themselves may contribute.

Our gratitude and esteem to all those who have brought us these mothers and daughters—opened the vistas, enabled the seeking eye.

I learned from the age of two or three that any room in our house, at any time of day, was there to read in, or to be read to. My mother read to me. She'd read to me in the big bedroom in the mornings, when we were in her rocker together, which ticked in rhythm as we rocked, as though we had a cricket accompanying the story. She'd read to me in the diningroom on winter afternoons in front of the coal fire, with our cuckoo clock ending the story with "Cuckoo," and at night when I'd got in my own bed. I must have given her no peace. Sometimes she'd read to me in the kitchen while she sat churning, and the churning sobbed along with *any* story. . . . She could still recite [the poems in McGuffey's Readers] in full when she was lying helpless and nearly blind, in her bed, an old lady. Reciting, her voice took on resonance and firmness, it rang with the old fervor, with ferocity even. She was teaching me one more, almost her last, lesson: emotions do not grow old. I knew that I would feel as she did, and I do.

EUDORA WELTY

19 Laura McPhee, *Untitled*, Ringoes, New Jersey, 1984

The Fourth of July is a special holiday in the black community, a freedom day for black people, when they all head south, head home. I was raised in Baltimore and on the Fourth, our family—mothers and grandmothers, fathers and grandfathers, children and aunts and uncles and cousins, all together—would sit down to crabs and watermelon and fried chicken. It's a different thing for white folks—they go out to the beach. This Fourth was doubly special: the first time blacks and Indians had gotten together in that area. Dr. King's oldest son was there, and it was the Bicentennial, a Happy Birthday to America. In the crowd I saw them: church folk, her husband a minister—good people—with their red, white and blue dresses made special for that day. I said, "My goodness!" and took their picture.

ROLAND FREEMAN

Roland Freeman, *Nellie G. Morgan and Tammie Pruitt Morgan, Bicentennial Celebration*, Philadelphia, Mississippi, 1976

21 Tom Bamberger, *Sophie with her Mother and Father*, Milwaukee, Wisconsin, 1984

 Leon Borensztein, *Mother with Daughter in Black Dress*, Santa Rosa, California, 1980

23 Joyce Baronio, *Portrait from 42nd Street Studio*, New York, New York, 1979

My dear, what you said was one thing
but what you sang was another, sweetly
subversive and dark as blackberries
and I became the daughter of your dream.

This body is your body, ashes now
and roses, but alive in my eyes, my breasts,
my throat, my thighs. You run in me
a tang of salt in the creek waters of my blood,

you sing in my mind like wine. What you
did not dare in your life you dare in mine.

MARGE PIERCY

Leslie was in her last six weeks of pregnancy. She had a two-year old at home, a twelve-year old daughter, Jenny, and an unpredictable income. At the time of the photograph, I, too, was pregnant with my third child and we shared those universal feelings of ambivalence, despair and determination. At first I photographed her lying on her back with a mattress in the yard. "Leslie," I complained "you just don't look big enough." "Big enough!" she cried and lumbered to her feet, yanking her dress up indignantly. Anyone who has been pregnant can appreciate her sentiments—and also the utter resignation and misery of that last huge month. Laughing, her daughter Jenny came over to her and we took the picture. Leslie delivered with ease a ten-pound baby at home that fall. SALLY MANN

 Sally Mann, *Jenny and Her Mother*, Lexington, Virginia, 1984

 Mary Kalergis, *The Surprise*, Virginia Beach, Virginia, 1985

When I was a young girl I wanted ten children—before I realized the difficulty in running a household full of small children. The housework and the financial burden are the most frustrating parts of mothering . . . . I wasn't prepared for how physically demanding it is; that I'd be so overwhelmed I couldn't even read a book. Strangely, though, this is probably the happiest time of my life . . . . The love you give your children is like black paper: it absorbs and you can't see it, but you know it's there. *from an interview by* MARY KALERGIS

 Mary Kalergis, *Laundry Day*, Charlottesville, Virginia, 1985

 Linda Brooks, *Mom at 55, me nearing 30*, Tamarac, Florida, 1981

29 Judith Black, *Laura and Self*, Cambridge, Massachusetts, 1984

The following spring, after her fortieth birthday, Rosie Vincent gave birth for the eighth time. It was a girl, Miranda Rose. Everyone was excited; there hadn't been a baby in the house for years.

Mum sat up in bed in her pretty nightgown, the pillows behind her bordered in *fleurs-de-lis,* holding her new treasure. Everyone hovered around, knocking against the dust ruffle, lying diagonally at her feet. Mum gazed into the infant eyes, seeing their strange clarity. She touched the tiny nose. She uncurled the fiddlehead fists and showed them to everyone lolling around. "You see?" she said. "Her father's hands exactly."

Then came the feeding. They watched her unbutton the nightgown and feel inside for the bosom. After fixing it to the baby mouth, and satisfied with it, she looked up. Caitlin and Sophie saw it—that wild look—only this time there was something added. It was aimed at them and it said: There is nothing in the world compares with this.

The eye was fierce. The baby stayed fast. There is nothing so thrilling as this. Nothing.

SUSAN MINOT

Images of myself as a full moon, full grown woman, and sky goddess began to appear in my work around 1976–1977. I enjoyed being pregnant, but needed time to clarify the many emotions and the physical experiences of these months. It was a time filled with magic, a feeling of giving and great pride in my body. These thoughts were on my mind at the time and happily my daughter was born the next spring. . . . BEA NETTLES

 Bea Nettles, *The Kiss*, Rochester, New York, 1979

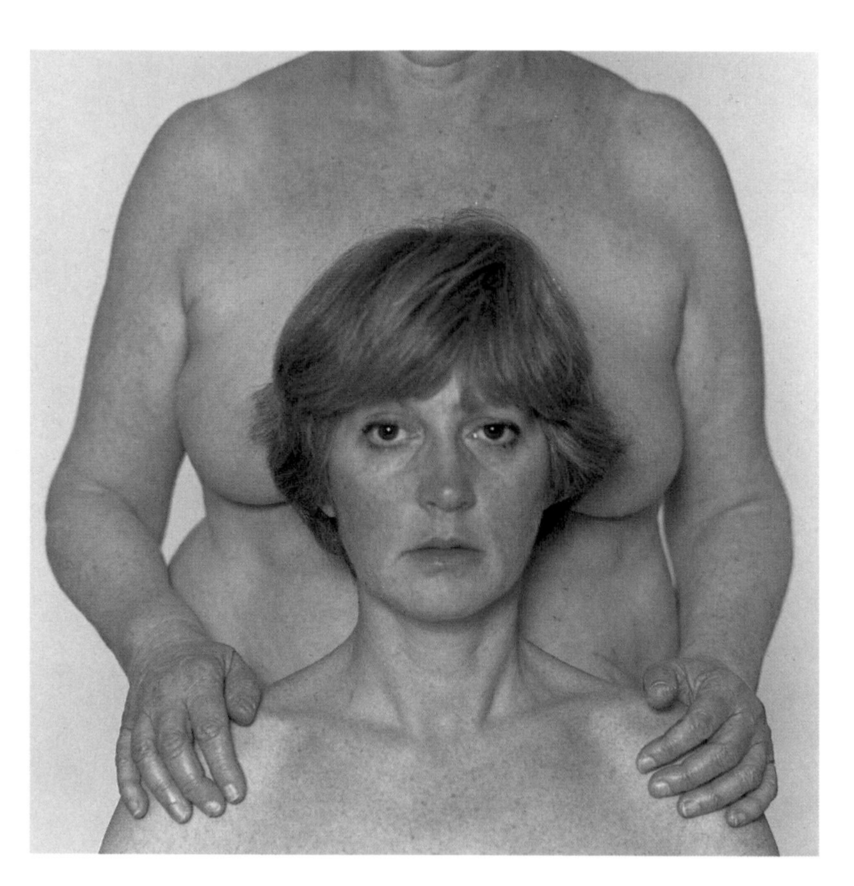

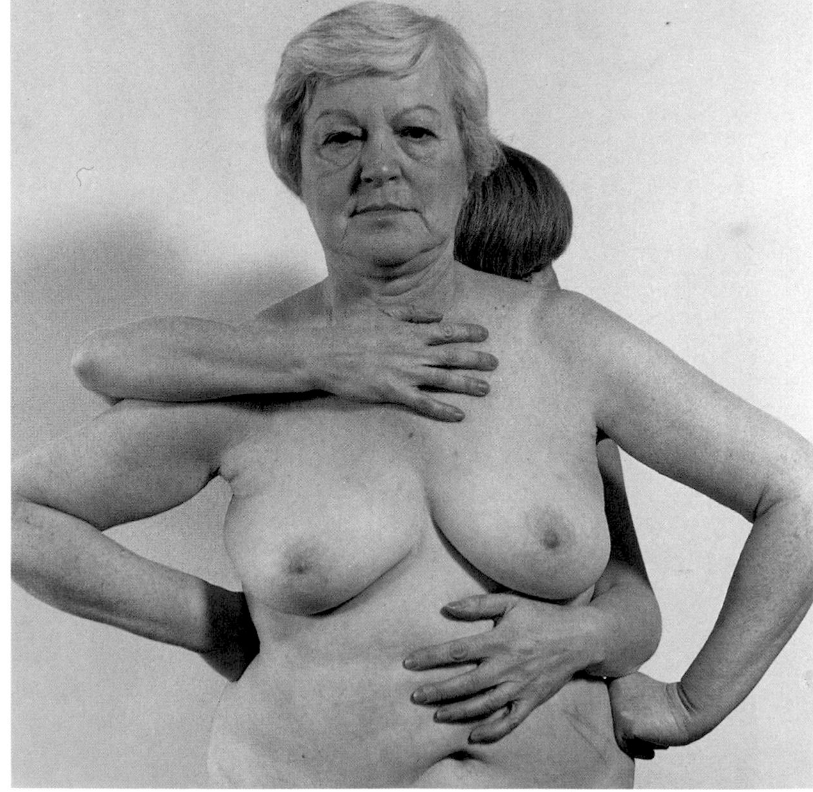

 Niki Berg, *Self-portrait with Mother (I, II)*, New York, New York, 1982

Susan Jahoda, from *Family Picture: Works Exploring Relationships Within the Group,* Berkeley Heights, New Jersey, 1985

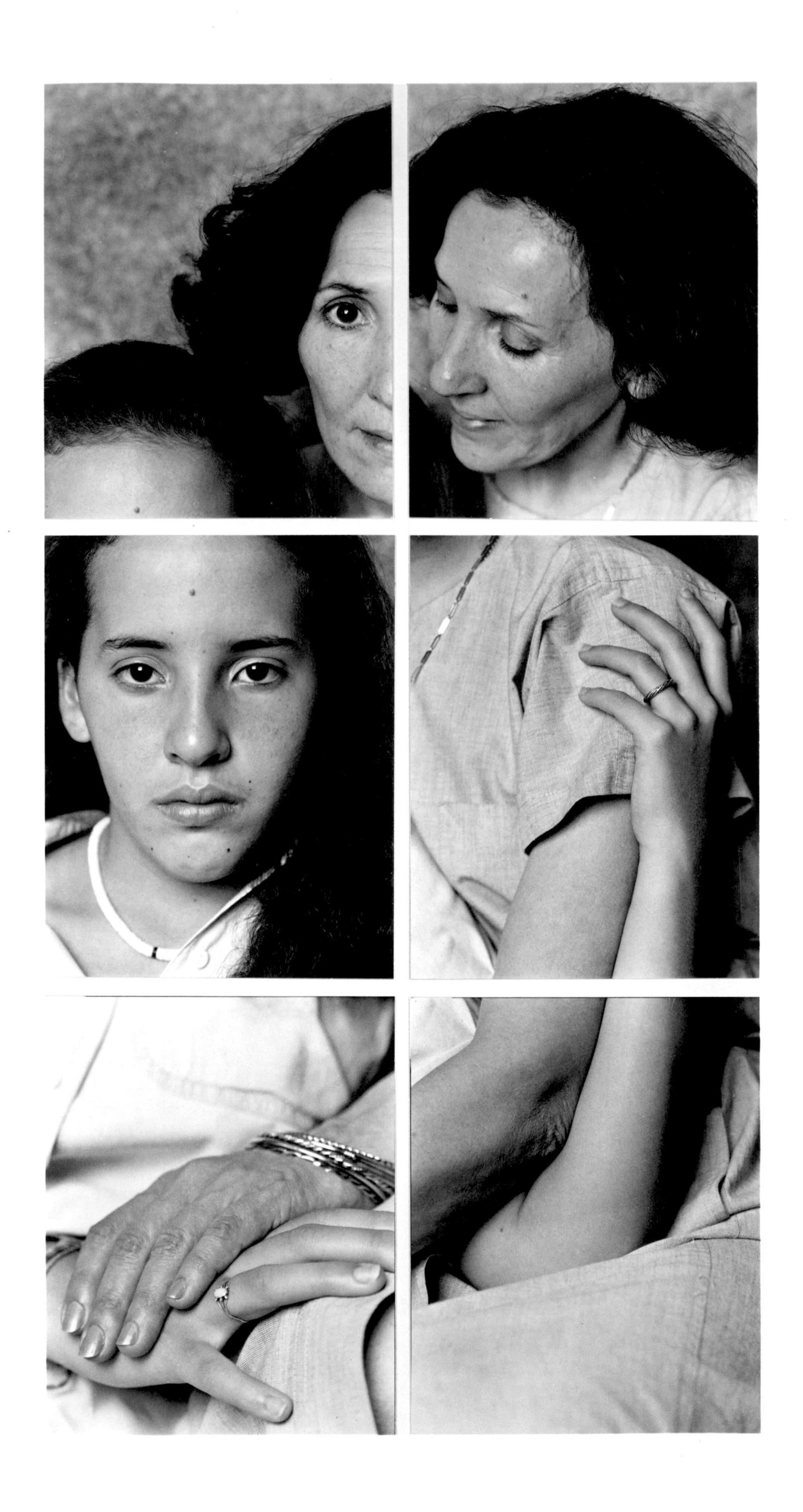

 Mark Berghash, *Melissa and Miriam*, New York, New York, 1986

 Barbara Crane, from *Chicago Beaches and Parks,* Chicago, Illinois, 1972-1978

Each time I order her to go
for a ruler and face her small
grubby outstretched palm
i feel before hitting it
the sting in my own
and become my mother
preparing to chastise me
on a gloomy Saturday afternoon
long ago, and glaring down into my own sad
and grieving face i forgive myself
for whatever crime i may
have done. as i wish i could always
forgive myself
then as now.

ALICE WALKER

37 Mark Goodman, *Untitled*, Millerton, New York, 1978

 Sage Sohier, *Rendezvous Beach*, Bear Lake, Utah, 1985

 Rosalind Solomon, *Untitled*, Washington, D.C., 1979

 Arthur Tress, *Mother and Daughter*, Boone, North Carolina, 1968

 Bruce Horowitz, *Untitled*, Wilmington, Delaware, 1983

MOTHERS

the last time i was home
to see my mother we kissed
exchanged pleasantries
and unpleasantries pulled a warm
comforting silence around
us and read separate books.

i remember the first time
i consciously saw her
we were living in a three room
apartment on burns avenue

mommy always sat in the dark
i don't know how i knew that but she did

that night i stumbled into the kitchen
maybe because i've always been
a night person or perhaps because i had wet
the bed
she was sitting in a chair
the room was bathed in moonlight diffused through
those thousands of panes landlords who rented
to people with children were prone to put in windows

she may have been smoking but maybe not
her hair was three-quarters her height
which made me a strong believer in the samson myth
and very black

i'm sure i just hung there by the door
i remember thinking: what a beautiful lady

she was very deliberately waiting
perhaps for my father to come home
from his night job or maybe for a dream
that had promised to come by
"come here" she said "i'll teach you
a poem: *i see the moon*
            the moon sees me
            god bless the moon
            and god bless me"
i taught it to my son
who recited it for her
just to say we must learn
to bear the pleasures
as we have borne the pain

NIKKI GIOVANNI

There was a spell cast in the room, as if time stood still at the eighteenth-century meeting house. Light streamed in, a benediction after the silent meeting, the telling of stories for the children. As if enacting an age-old scene, the child sought out her mother for comfort. It was the kind of moment that will come once, and never again . . . . FRANCES McLAUGHLIN-GILL

 Frances McLaughlin-Gill, *The Listeners, Quaker Meeting House*, Jericho, New York, 1967

The parish priest had called and explained I was on an assignment for a Catholic magazine, a sort of messenger of the Lord, and would they agree to have me photograph their daughter before her First Communion. They were simple Catholic people, and there was a reverence of simplicity and profundity apparent in their belief. The little girl was so self-conscious, but pleased with herself; and later, at the moment of her First Communion, she was perfectly contained in her fear.

LARRY FINK

 Larry Fink, *First Communion*, Bronx, New York, 1961

45 Larry Fink, *Henrika Frajlich*, New York, New York, 1984

 Charles Harbutt, *Untitled,* Brooklyn, New York, 1963

47 Hope Wurmfield, *Jean and Dria*, New York, New York, 1982

TRICKS

My mother
the magician
can make eggs
appear in her hand.
My ovaries
appear in her hand, black as figs and
wrinkled as fingers on wash-day.

She closes her hand
and when she opens it
nothing.

She pulls silk scarves put of her ears
in all colors, jewels from her mouth,
milk from her nipples. My mother the naked
magician stands on the white stage
and pulls her tricks.

She takes out her eyes
The holes of her sockets
fill with oil, it seeps up,
with bourbon and feces.
Out of her nostrils
she pulls scrolls
and they take fire. . . .

In the grand finale
she draws my father
slowly out of her cunt and puts him
into a tall silk hat
and he disappears.

I say she can turn anything
into nothing, she's a hole in space,
she's the tops, the best
magician. All this

I have pulled out of my mouth right
before your eyes.

SHARON OLDS

Raisa Fastman, *Teal Costigan with Her Mother, Landis Costigan*, San Francisco, California, 1985

Several years ago a friend and I were discussing the various aspects of being our mothers' daughters, the similarities we each bore to our mothers, difficult to accept because we were both—shall I confess?—desperate for our own identities. This recognition launched a project involving seventy mothers and daughters. Some turned it down: because of recent arguments, long-standing silences, tenuous emotional connections or fear. One mother set three separate dates before the planets were astrologically satisfactory. Another drank a six-pack in bed before she gave permission. Though there is no photograph that can describe even the most tranquil flesh and blood relationship . . . the point is to look and consider the human possibilities.

CARLA WEBER

 Carla Weber, *Donna and Diane*, Santa Fe, New Mexico, 1982

 Carla Weber, *Tang Chung, Lisa Lu, Lucia and Loretta*, Los Angeles, California, 1986

 Abigail Heyman, *Untitled*, New York, New York, 1984

 Randy Matusow, from *Unbroken Bonds* series, Queens, New York, 1985

 Nicholas Nixon, *Untitled*, Hyde Park, Massachusetts, 1979

 Mary Kalergis, *Untitled,* Brooklyn, New York, 1985

 Jason Laure, *Woodstock Music Festival*, Woodstock, New York, 1969

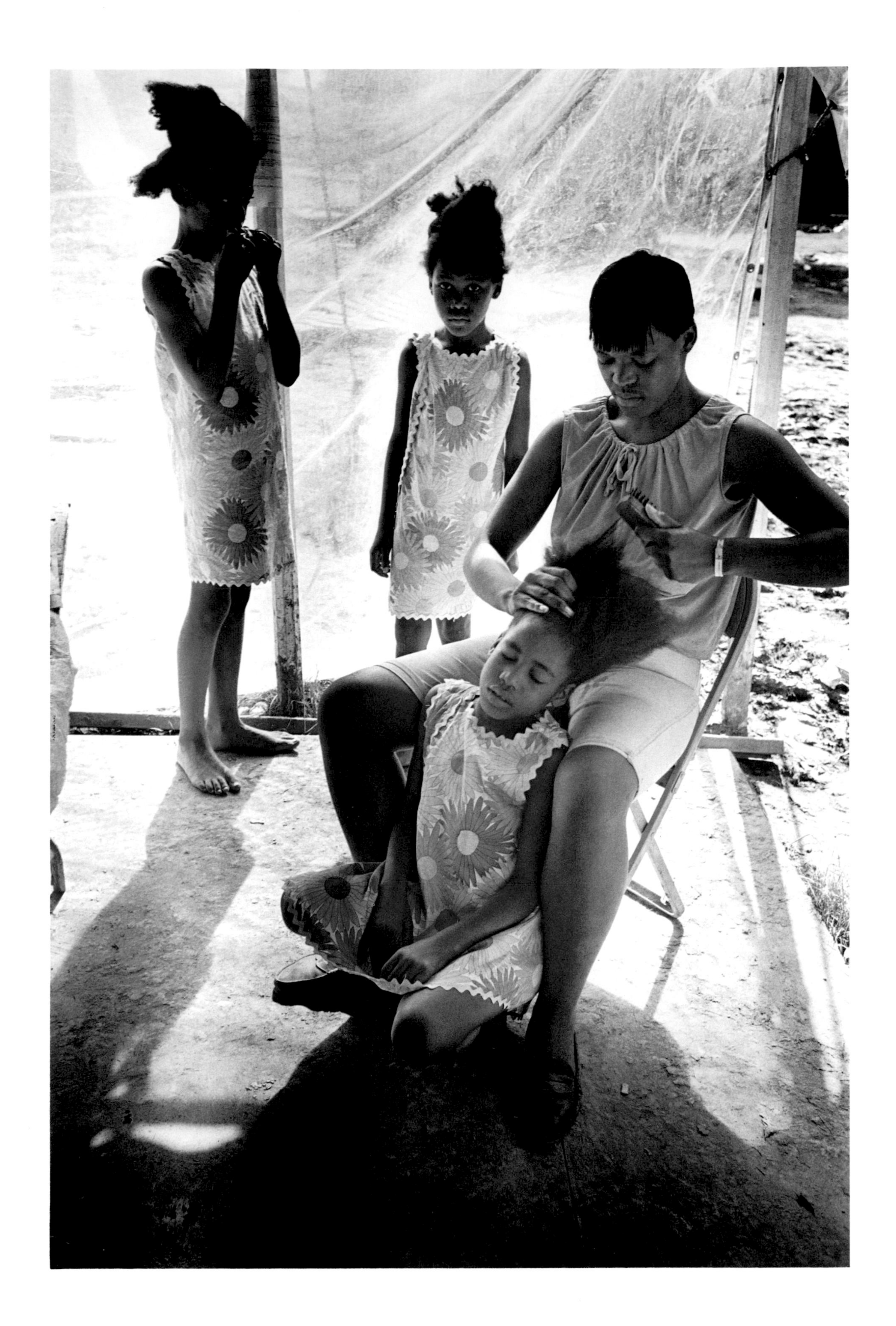

 Jill Freedman, *Hair*, Resurrection City, Washington, D.C., 1968

I am her only novel.
The plot is melodramatic,
hot lovers leap out of
thickets, it makes you cry
a lot, in between the revolutionary
heroics and making good
home-cooked soup.
Understand: I am my mother's
novel daughter: I
have my duty to perform.

MARGE PIERCY

 Danny Lyon, *August*, New Mexico, 1979

This is hill country, rocky, with thin, poor soil: a hamlet called George's Branch, in Hidden Valley, home to twenty families. The odds of living here and having any other life, of improving their lot, is nonexistent. This woman could only bequeath to her children a legacy of hardship, but one became aware of the intensity of their feeling for the land, their bond with their homeplace that sustained them. PAUL FUSCO

 Paul Fusco, *Appalachian Farm Family*, Harlan County, Kentucky, 1965

The Cocopah Indians are the smallest Indian tribe in the United States, living on a piece of land you can walk across in a half an hour, land that is just desert, dust, and weeds. They gather sticks of wood for a living. Their water, from a creek off the Colorado, is polluted and brackish. But they share this feeling for the land; they know, these mothers, that it will be tough on their kids. Surely there must be some way for society, this society that is so rich, to honor their need? PAUL FUSCO

 Paul Fusco, *Cocopah Indian Family*, Cocopah Reservation, Arizona, 1970

One moment she looks fresh and beautiful; the next she looks beaten out, whipped by life. She is twenty-nine, with opalescent eyes, her skin is the color of sand. Her hands are at once gnarled and delicate. She has a deep hill-country voice. Sometimes she sings along with Moe and Joe on the tape deck and you can see it's the purest pleasure she knows. Her favorite song is "Your Cheatin' Heart." HERMAN LEROY EMMET

 Michal Heron, *Navaho Family*, Mexican Water, Arizona, 1981

63 Herman Leroy Emmet, *Fruit Tramps: The Tindal Family*, Edneyville, North Carolina, 1979

DAUGHTERLY

So many women, writing,
escaped their mothers—
mine, in her nightgown,
retreated into a wordless depression.

If only I could have
spoken for her,
but she turned her face to stone,
her curly hair to snakes,

and her tongue dried up
trying to escape her children.
We skated her surface,
the old ice pond

with its treacherous
depths and greenish patches.
When she melted for moments
she touched my cheek, snowflake,

like a hot penny
burning a hole to my heart.
In her inward chill
I seared myself over,

a young girl
skating away, writing
on air
with a red muffler.

Her silence; my silence:
the house, its stubborn necessities;
the snow, her scabbed
depression, drifted secretly

till even the blades of our
ice skates stopped
their thin persistent scraping on
her winter; the knife sharp air.

KATHLEEN SPIVACK

A photograph is a sort of daughter: conceived one hopes (but not necessarily) in rapture, it must come to life on its own. One's relationship to it in the beginning largely consists of carting it around and having hopes for it. And *if* it is a successful one, it develops a personality, goes off alone, leads a life of which its mother might not be the best interpreter.

SAGE SOHIER

Sage Sohier, *Untitled*, Brookline, Massachusetts, 1986

 Eudora Welty, *Spanking*, Hinds County, Mississippi, ca. 1934

67 Curt Richter, *Patty and Mahala,* Garrison, New York, 1982

 Joan Liftin, *Untitled*, Upstate New York, 1974

69 Elaine O'Neil, *Self-portrait With Daughter*, Dorchester, Massachusetts, 1985

A DAUGHTER (1)

When she was in the stranger's house—
    good strangers, almost relatives, good house,
    so familiar, known for twenty years,
    its every sound at once, and without thought,
    interpreted:
                        but alien, deeply alien—
    when she was there last week, part of her wanted
    only to leave. It said, *I must escape*—no,
    crudely, in the vernacular: *I gotta get outta here,*
    it said.

And part of her
ached for her mother's pain,
her dying here—at home, yet far away from home,
thousands of miles of earth and sea, and ninety years
from her roots. The daughter's one happiness
during the brief visit that might be her last
(no, last but one: of course there could always be
what had stood for years at the end of some highway of
factual knowledge, a terminal wall;
there would be words to deal with: funeral, burial,
                                                        disposal of effects;
the books to pack up)—her one happiness this time
was to water her mother's treasured, fenced-in garden,
a Welsh oasis where she remembers adobe rubble
two decades ago. Will her mother now
ever rise from bed, walk out of her room,
                                                            see if her yellow rose
has bloomed again?
Rainbows, the dark earthfragrance, the whisper of
                                                                    arched spray:
the pleasure goes back
to the London garden, forty, fifty years ago,
her mother, younger than she is now.
And back in the north, watering the blue ajuga
                              (far from beginnings too; but it's a place
                              she's chosen as home)
the daughter knows
another, hidden, part of her longed—or longs—
for her mother to be her mother again,
consoling, judging, forgiving,
whose arms were once
                                        strong to hold her and rock her, . . .

DENISE LEVERTOV

71 Joseph Szabo, *The Wedding Bouquet*, Long Island, New York, 1977

She is a single mother raising two daughters, a neighbor of mine. They did this skit in their backyard for the family and for friends, complete with a lasso number and a totem finale. The relationship is a very close one between them, even now that the girls are almost grown. RAISA FASTMAN

 Raisa Fastman, *Wonder Woman Skit*, Albany, California, 1985

 Susan Copen-Oken, *Coach (The Bell Family)*, Orlando, Florida, 1985

 Elliott Erwitt, *Karla Cohn and Her Mother*, Armonk, New York, 1959

75 Charles Harbutt, *Teenage Marriage*, Granite City, Illinois, 1965

One night we crashed a wedding dinner,
but not the guests. We crashed the chef.
We put dollar bills in the salad, right beside
the lettuce and tomatoes. Our salad was a winner.
The guests kept picking out the bucks, such tiny thefts,
and cawing and laughing like seagulls at their landslide.
There was a strange power to it. Power in that lovely paper.
The bride and groom were proud. I call it my Buck Wedding Caper.

My own ideas are a curse for a king and queen.
I'm a wound without blood, a car without gasoline
unless I can shake myself free of my dog, my flag,
of my desk, my mind. I find life a bit of a drag.
Not always, mind you. Usually I'm like my frying pan—
useful, graceful, sturdy and with no caper, no plan.

ANNE SEXTON

My husband, Pat, has a theory about watering our newly seeded lawn. The water has to trickle from heaven and fall like tender little rain drops . . . otherwise the lawn won't grow properly.

from an interview by BILL OWENS

 Bill Owens, from *Suburbia*, Vesta San Ramone, California, 1972

I was walking along the beach and saw this group: mothers and daughters, sisters and sisters, women together, having a very good time.
ARLENE GOTTFRIED

 Arlene Gottfried, *Summer*, Coney Island, New York, 1976

 Margaret Randall, *Mother and Daughter Riding*, Granger, Washington, 1984

That I not be a restless ghost
Who haunts your footsteps as they pass
Beyond the point where you have left
Me standing in the newsprung grass,

You must be free to take a path
Whose end I feel no need to know,
No irking fever to be sure
You went where I would have you go. . . .

So you can go without regret
Away from this familiar land
Leaving your kiss upon my hair
And all the future in your hands.

MARGARET MEAD

Trellis and Aneta had a fight. It started out about the wine-red dress. Trellis said the dress was too small. Aneta said it wasn't. Trellis said the weather was too warm for velvet. Aneta said it wasn't. She loved the dress and was unhappy about growing out of things.

Then they started arguing about her hair. Trellis wanted to style it into the usual "long curls." Aneta wanted it brushed out. She cried and stomped and ripped the red dress under the arm. Trellis said that showed it was too small.

Aneta went to the bathroom and brushed out her hair. She couldn't really get it to go a way she liked. She wanted to be pretty, but not in the way Trellis wanted. She wanted to be pretty without curlers and hot curling irons. She didn't understand why it should have to be painful to be presentable.

Aneta W. Sperber 1978

Though it was made in the present, it has to do with a specific remembered incident, a mother teaching her daughter how to fix her hair, teaching her through this presentation of self about time, about the roles given out to women in society, about a wider social and political context. This is one portrayal of the relationships of women to families, one aspect of what women do, what women are. ANETA SPERBER

Aneta, Bloomington, Indiana, and Montreal, Quebec, 1978

Dear Mom
I love you,
and I would like
to be friendsagun.

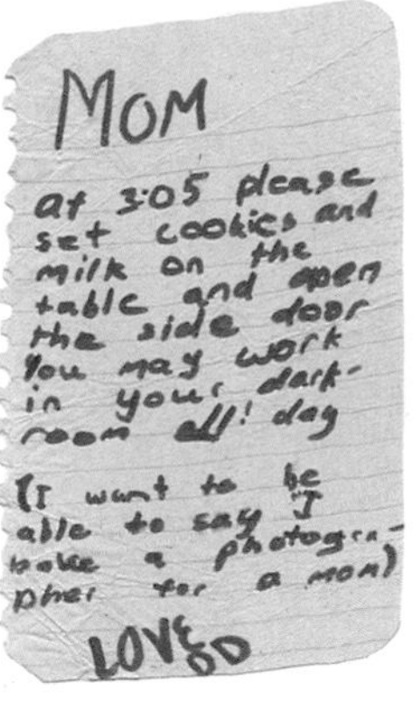

MOM

at 3:05 please
set cookies and
milk on the
table and open
the side door
You may work
in your dark-
room all! day

(I want to be
able to say "I
have a photogra-
pher for a mom)

LOVE OD

Rosemary Porter, *"I'd like to be friends again . . ."*, Boston, Massachusetts, 1980
Rosemary Porter, *"A photographer for a Mom . . ."*
(photographed by Judith Sedwick), Boston, Massachusetts, 1980

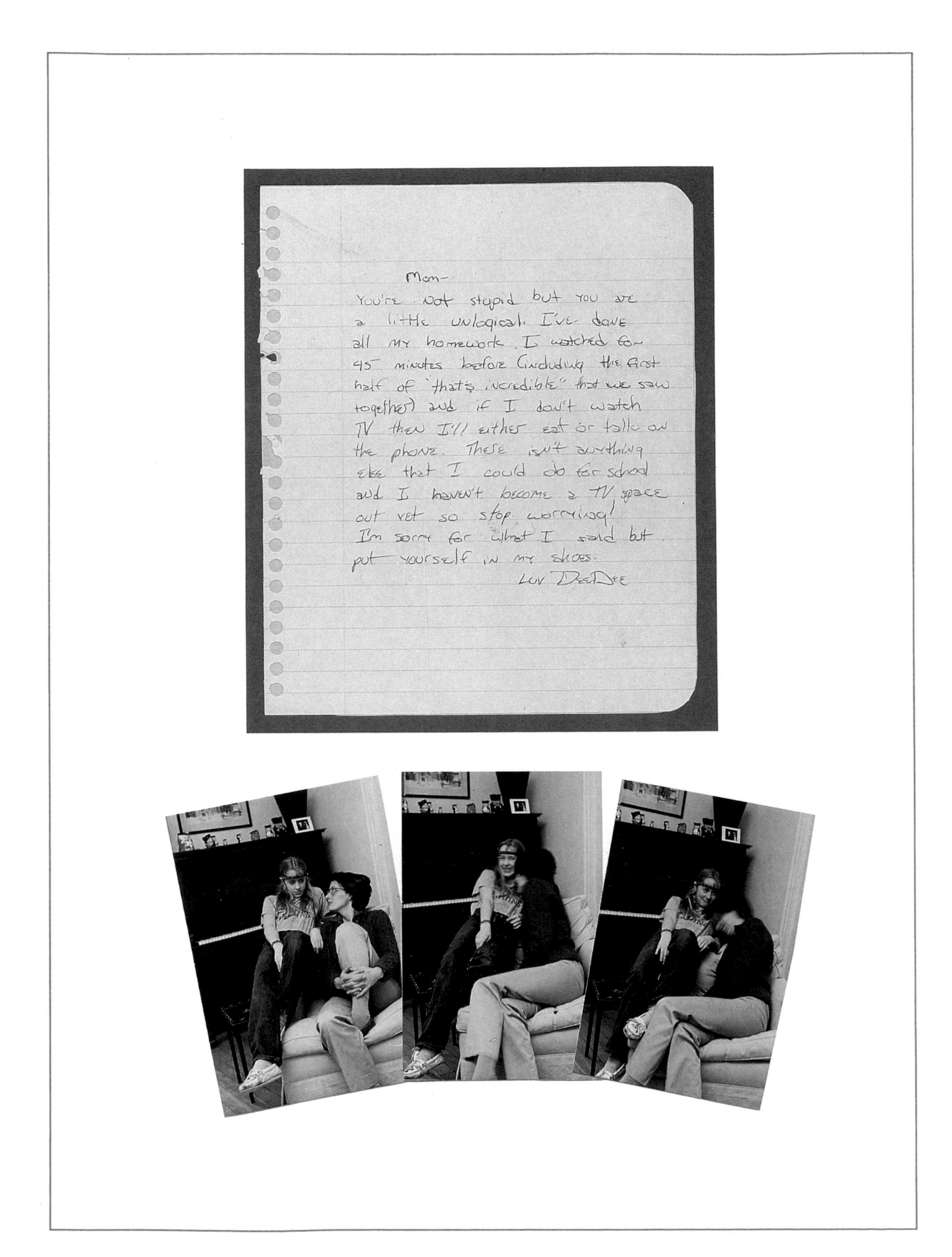

Mom–
You're not stupid but you are
a little unlogical. I've done
all my homework. I watched for
45 minutes before (including the first
half of "that's incredible" that we saw
together) and if I don't watch
TV then I'll either eat or talk on
the phone. There isn't anything
else that I could do for school
and I haven't become a TV space
out yet so stop worrying!
I'm sorry for what I said but
put yourself in my shoes.
Luv DeeDee

Rosemary Porter, *"You're not stupid but . . ."*
(Photographed by Judith Sedwick), Boston, Massachusetts, 1986

While working on a series of pictures of coal miners I met this woman, a single parent from a coal mining family whose father and grandfather were miners. In that area, the only choice she had was to be a maid in a hotel or go into the mines. She chose to be a miner. Here she is with her daughter, after getting through her shift, and later at home.

MILTON ROGOVIN

 Milton Rogovin, *Mother and Daughter*, Beckley, West Virginia, 1981

For three years I worked in a six-block radius on a series of streets next to my optometric office in Buffalo, taking pictures of homes, churches, businesses and people. Her daughter lives with her grandmother now, but came back for the second picture, taken thirteen years after the first.

MILTON ROGOVIN

 Milton Rogovin, *Mother and Daughter*, Buffalo, New York, 1973 and 1985

 David Graham, *Three Generations*, Louisville, Kentucky, 1983

 Todd Merrill, *Susan and Blaze*, San Francisco, California, 1986

She, who had no worldly goods to leave, yet left to me an inexhaustible legacy. Inherent in it, this heritage of summoning resources to make—out of song, food, warmth, expressions of human love—courage, hope, resistance, belief; this vision of universality, before the lessenings, harms, divisions of the world are visited on it.

She sheltered and carried that belief, that wisdom—as she sheltered and carried us, and others—throughout a lifetime lived in a world whose season was, as still it is, a time of winter.

TILLIE OLSEN

 Todd Weinstein, *Mother and Child*, New York, New York, 1980

 Eric Breitenbach, *Untitled 1 and 2*, Edgewater, Florida, 1986

When Clementine was born, I started taking pictures of her—this one was taken when she was three or four days old—and then, perhaps because I finally figured out how to take baby pictures, I stopped my other work and have done little else since.

NICHOLAS NIXON

 Nicholas Nixon, *Untitled*, Cambridge, Massachusetts, 1985

EXCLUSIVE
(for my daughter)

I lie on the beach, watching you
as you lie on the beach, memorizing you
against the time when you will not be with me:
your empurpled lips, swollen in the sun
and smooth as the inner lips of a shell;
your biscuit-gold skin, glazed and
faintly pitted, like the surface of a biscuit;
the serious knotted twine of your hair.
I have loved you instead of anyone else,
loved you as a way of loving no one else,
every separate grain of your body
building the god, as I built you within me,
a sealed world. What if from your lips
I had learned the love of other lips
from your starred, gummed lashes the love of
other lashes, from your shut, quivering
eyes the love of other eyes,
from your body the bodies,
from your life the lives?
Today I see it is there to be learned from you:
to love what I do not own.

SHARON OLDS

93 Jock Sturges, *Sandy, Catherine, and Angella Smith*, Block Island, Rhode Island, 1984

Rhondal McKinney, *Gussie (Edith) Klump and Her Daughters Lois Hines and Waunita Geshwiln*
from "Farm Family" series, McLean County, Illinois, 1985

They live in a county known for having the richest soil in the state and have spent their lives in that one spot. One daughter is a schoolteacher, the other works inside the home; they live within ten miles of their mother. You could tell that they spent an awful lot of time together, that they were a close threesome. They would come right into the house and do some little chore that needed doing, just like they were coming home after school. They were still Gussie's girls. What I liked was the ease with which they stood together. I'm a parent myself, though my children are young, and it seems to me that in most relationships between parents and children, all those years of separation, of growing up, take their toll, split people apart. But Gussie and her daughters were like a close knot—there was no separating them.

RHONDAL MCKINNEY

There are many photographs which do not contain the full truth, that do not reveal form, that do not show a coherence in its deepest sense . . . . Often, however, a composition is a major way for a photographer to show the wholeness of life.

ROBERT ADAMS

 Robert Adams, from *Our Lives and Our Children*, Rocky Flats, Colorado, 1983

97 Harry Callahan, *Untitled*, Chicago, Illinois, 1953

THE HEART OF THE INEFFABLE

Estelle Jussim

Mothers and daughters! Are there any fantasies or ideal images left to us in this tough-minded age of psychological realism? Do we still yearn (without admitting it) for the imagined paradise of lost Mother-Daughter delights that some of us have never known? Psychologists tell us that *Mother* is that all-giving, all-dominating, all-powerful figure from our fairy-tale infancies, the giant in the nursery. She is transmuted into the evil stepmother of those tales, yet simultaneously she is also the life-enhancing, sweetly tender goddess whose beauty, goodness, and self-sacrifice we could never hope to equal. Ultimately, she seems to contain all the mysteries of adulthood.

Daughter, on the other hand, simply connotes a biological relationship, a social role of no great consequence, and (depending on the culture into which a daughter is born) an expensive liability requiring dowries and protection. A society that devalues women in general may decide that the birth of a daughter is an unaffordable luxury, since she cannot become a warrior, a hunter, or a chief. Fortunately, most human groups no longer dispose of their daughters on the nearest hilltop. But, as we look at the pictures in this collection, we should keep a firm grasp on our history, for even in the relatively few decades we have enjoyed photography, the changing iconography that inevitably accompanies changing social attitudes is conspicuous and tangible.

It has been widely recognized that even the greatest portrait can capture only so much of an individual's personality and character, not all of that person's physical attributes, and certainly not a permanently ascribable mood. An attempt by a photographer to convey not only one, but two persons and their relationship, might seem to be exceedingly difficult, if not impossible. To portray two persons defined as mother and daughter is to define a relationship fraught with cultural and emotional overtones. Such intensity of meaning would seem to demand skillful decoding. Perhaps, also, it requires a grasp of visual language that not all of us possess. Even if we did possess such a visual language, it might prove to be so ethnocentric and tempocentric as to defy our desires for significant universal meanings. This collection makes no pretense of offering more than an intelligent sifting of contemporary imagery, which, upon examination, can reveal much about contemporary life and our implicit ideologies concerning motherhood.

Wynn Bullock once observed that photographs peer at life and attempt to record it truthfully, but can only transcribe two of the four of life's dimensions. Reduced to the plane of two dimensions on a flat piece of paper, constrained (usually) within the four straight edges of a rectangle, photographs lack two crucial factors: the sense of ongoing time in the three physical dimensions, and the unseen (because unseeable) fourth dimension Bullock called "spirit."

Even if they could encompass all the dimensions experienced by living human beings, photographs offer complexities of meanings, not single, precisely definable and verifiable meanings. Like value of any kind, meaning is ascribed by the viewer, the person examining the image in his or her own specific moment of time. We decode a picture the only way we can: through our visual enculturation, interpreting images by means of our idiosyncratic backgrounds as well, including socio-economic class, political bias, educational level, religious affiliation or spiritual inclination, competence with symbolism and other aspects of iconography, and a multitude of other vital influences. Of these, perhaps none is more important than how we relate to our own status and history in the hierarchy of family relationships.

What is remarkable, then, is that creative photographers manage to convey a great deal about reality, about life, about the spirit of human beings, and about superordinate realities sometimes called the *Zeitgeist*, or the Soul of an Era. Photographs, which always require close study and attention before their various potential meanings can reveal themselves, can and do provide a substantial number of clues about relationships. We simply have to be tuned in to those clues and to have both patience and humility in assigning their significance. While verbal information often proves crucially useful in this enterprise, it is not always as helpful as we might suspect.

Perhaps surprisingly, the most explicit caption may fail to explain the specifics of either a relationship or a situation depicted in a single photograph. Consider, for example, the Mathew Brady albumen print which the Library of Congress identified as *Rose O'Neal Greenhow with her daughter in the courtyard of the Old Capitol Prison, Washington, D.C.* (page 99, left) The caption for this 1865 picture seems explicit enough, yet it tells us nothing about the fact that Rose Greenhow was a Confederate spy captured by Union forces. That fact explains why she was photographed in the courtyard of a prison, but does not explain the presence of her daughter, who clutches her tight-lipped mother with a strong, protective embrace. Are they about to be separated? Is the mother about to be hanged? How did Rose Greenhow manage to maintain her elegant costume, her dainty gloves, her air of nobility? Without a considerable amount of biographical research, we may never know what, exactly, is being depicted here. Yet perhaps we can manage to decipher the mood of strained affection and taint of sorrow. We can do this because we tend to believe that emotions reveal themselves through the face, body posture, gestures of

(left) Matthew Brady *Rose O'Neal Greenhow with her daughter in the courtyard of the Old Capitol prison,* Washington, D.C. 1865

Henry Peach Robinson, *Fading Away*, 1858

hands, just as we interpret the historical moment and social class of persons through their apparel.

American audiences and American photographers of the mid-nineteenth century cared much more for facts than for the melodrama of an "art" picture by the British photographer Henry Peach Robinson. *Fading Away* (page 99, right) is obviously a fiction, a tableau, so we do not expect realistic documentation, but try to interpret the symbols provided by this 1858 concoction. Watched over by an older sister (or perhaps a good friend), a young woman fades away to die amid the mawkish trappings of a scene that could have come straight out of Dickens. The bonneted mother seems resigned to her child's fate; she holds a book, undoubtedly moral teachings about accepting death. Compiled from a number of separate negatives, *Fading Away*, with its powerful symmetry, artificial poses, and distraught male figure (the father? a husband?) bids us all too dramatically to weep. The bright sky, however, offers hope of heaven: the daughter will soon be among the angels. Too reminiscent of theater to be taken seriously today, the picture tells little about the grief of the mother. From a knowledge of Victorian conventions, we know that the mother was taught Christian resignation, and to express grief most genteelly, without the rending of clothes, wailing, and uncontrollable anguish that might emanate from other ethnic groups. We should also take note that it was all too common in the mid-nineteenth century for mothers to lose their daughters (and sons) to the ravages of tuberculosis, pneumonia, or typhoid fever.

America's greatest portrayer of the mother-daughter relationship in the late nineteenth and early twentieth centuries was indisputably Gertrude Käsebier. In her justly famous photograph, *Blessed Art Thou Among Women* (page 100, left), an elegant mother seems to be urging her adored daughter to cross the threshold, a metaphor for entering adult life, and to pass confidently from one state of being to another. The lovely formal aspects of this 1899 gum platinum print were the result of Käsebier's expert posterization, a technique that emanated from both the teachings of Arthur Wesley Dow and *japonisme* in artistic posters themselves. The decorative darks concentrated in the figure of the daughter, with her precise contour, are enhanced by contrast to the wraithlike whiteness of the mother, a woman who by her overtly, loving connection with the daughter epitomized what mothers were supposed to be. It was, as the author and critic Ann Dally has remarked, the era that "invented motherhood" as a relationship of an almost suffocating closeness between the maternal figure and the offspring. Yet Käsebier displays that intimacy not as suffocating, but as sublimely supportive.

Nothing here presages the imminent death of this beautiful child. These were real people, not the stagey characters of *Fading Away*. The pair were Käsebier's friend, Frances Lee, and her daughter. In a later Käsebier photograph of 1900, called *The Heritage of Motherhood* (page 100, right), Frances Lee was transfigured into a universal symbol of grieving motherhood, with her agony undiminished by the false piety and repressed resignation of the Robinson pastiche. Käsebier took her subject seriously. She was one of the first photographers who rendered the mother-daughter relationship without tear-jerking sentimentality. She granted both female roles respect and admiration, a remarkable accomplishment at a time when to be considered "blessed among women" because one had a daughter was possibly a rare proposition in an American culture that worshipped male progeny. (If women were still on a pedestal in 1900, it was because in that rather awkward position they were hardly able to compete with men for power or even the dignity of the vote.)

The poet Adrienne Rich encapsulated the problem: *how we dwelt in two worlds / the daughters and the mothers / in the kingdom*

(left) Gertrude Käsebier, *Blessed Art Thou*, 1899

Gertrude Käsebier; *The Heritage of Motherhood*, 1900

of the sons. Iconographically, the worship of a son rather than a daughter has been one of the most enduring and influential images in Western art since the establishment of the Virgin Mary as a primary figure of worship. In our unconscious responses to images of mothers with children, it would be difficult to ignore the residue of all the thousands of reproductions we have seen of paintings showing the Virgin Mother and her God-son. That is perhaps why Lewis Hine's *Ellis Island Madonna* (page 101, left) seems strained and falsely titled. The immigrant mother's adoration is appropriate for the traditional idea of a Madonna, but the daughter she holds so firmly gazes upward with a most peculiar stare, and not at the mother, who is behind the child. Taken with the explosive light of flash powder, the picture surprises because the child did not wince. Flash powder was fast, but tended to make nervous wrecks out of unsuspecting subjects. That a mother was holding a child seems an insufficient excuse to dignify this picture with the resonance of one of the greatest themes of Christian art. Besides, it should have been obvious to Hine that no Madonna had ever worshipped God immanent in a girl-child.

Social class as well as physical presence are two clues to the beauty and charm of James Van der Zee's portrait of a mother with two daughters. Against a typical studio backdrop of the 1930s, the trio radiates a quiet beauty (page 101, right). The grace and poise of these female members of the black bourgeoisie were heightened by the self-aware prettiness of the little girl who has wrapped her mother in two arms. Not quite so adorable, the other daughter is more relaxed. The mother could easily have posed for Raphael, and the group is the visual antithesis of the farm wife with children so typical of the work of Michael Disfarmer (page 102, left).

In a frontal pose that bluntly presents an enraged daughter and a gently resigned mother, this Disfarmer studio portrait is completely bare of environmental or fictional ambience, a technique that Richard Avedon adopted for its ability to concentrate total attention on the participants in a photograph. The mother's willingness to be recorded in her gaudy cheap dress and anklets with sandals indicates not a lack of aesthetic taste but of education, and deep poverty. The mother restrains her glaring daughter with a protective hand that has known labor. For the sake of historical accuracy, it should be noted that it is sometimes impossible to distinguish very young girls from boys because in many areas it was customary for both to wear dresses in infancy. But this sturdy youngster seems unquestionably a girl.

In another Disfarmer double portrait (page 102, right), one which features a wonderful rhythm of diagonally placed forearms and vertical legs, a rural mother perches with her teenage daughter in an affectionate and casual pose. The young girl is comfortable, relaxed, and intelligent, and her mother, looking away from the camera, seems as solid and strong as our stereotype of farm wives and pioneer women would have us believe. No resignation, no anger here; these are straightforward people, so real one feels able to strike up a conversation as soon as they stop posing. If farm life was hard in the thirties and forties, nothing in this Disfarmer image reminds us of the terrifying Great Depression that echoes so implacably in the work of Dorothea Lange.

Migrant Mother, Nipomo, California, 1936 (page 103) is Lange's most famous picture, one that is almost an embarrassment to reproduce yet another time, as the concerned mother survived the dust bowl agony to tell the world that she had never received a penny for the thousands of times this picture has been printed. Certainly, she was immortalized as the very essence of the infinite cares of motherhood. How to feed, clean, clothe these daughters in the midst of social disintegration? How to summon the strength to go on working as a grossly underpaid pea-picker in the hell of a California migrant

(left) Lewis Hine,
Ellis Island Madonna, 1905

James Van der Zee,
Sunday Morning, 1936

camp? How to maintain the unity of this desperate family? What would be the future of these starving daughters without the opportunity of an education? Would the father be able to find work? The Lange image, so sculptural and sure in its details, makes a direct impact on the emotions, reminding us of the anguish of extreme poverty before it invites us to ask questions about the fate of individuals confronted by social disaster.

While the Farm Security Administration photographers were out scouting for pictures that they hoped would arouse the sympathy of an America distracted by nationwide unemployment, even more dangerous threats were brewing in Europe and the Far East. World War II—that last "good" war where the alternatives to defeating Hitler's Germany were nonexistent—hurled destruction on millions of innocent civilians. After the war, Americans turned to "togetherness" and family life with considerable enthusiasm. In Eve Arnold's interior scene (page 104), we find a sailor home on leave, his wife continuing the middle-class niceties of white gloves for the dainty blonde daughter. The youngest plays happily with the Sunday funnies, oblivious both to the socialization process being inflicted on her sister and to the meaning of her father's uniform. Perhaps the older daughter is being readied for Sunday school, or the family has just returned from an early visit to church (the clock reads 10:55 and the light indicates that this is morning). The history of the movies tells us that the dolled-up child has been coiffed and dressed in imitation of that great culture icon of the 1930's, Shirley Temple, whose singing, dancing, and unbridled cuteness won the hearts of many a mother. The artless *accoutrements* of the room include an early television set (television did not become ubiquitous until the 1950's), which, marvelously, mirrors back the daughter who is already the mirror of a cultural ideal. That the mother should be the transmitter of this ideal is obvious; less obvious, perhaps, is that the daughter is without the power to choose an alternative vision of female childhood. Meanwhile, the young father buries himself in "man's work," reading the sports pages or the political news. Ordinary women were not expected to be interested in politics.

One of the most poignant mother-daughter images of the late 1940's was taken by Jerome Liebling at a Jewish wedding in Brooklyn (page 105). Here, side by side, viewed with Liebling's characteristic combination of brutal honesty and compassion, are two generations. It has been said that a man can tell the future of his intended bride by looking at the face and figure of his intended mother-in-law. Will this sensual daughter indeed grow to resemble the monumental mother? Together, they offer a visual contrast as stark as that between the delicacy of a Tanagra figurine and the baroque extravagance of late Hellenistic sculpture.

In the diary of Anaïs Nin, she describes her mother in terms that illuminate Liebling's portrait: "She was secretive about sex, yet florid, natural, warm, fond of eating, earthy in other ways. But she became a Mother, sexless, all maternity, a devouring maternity enveloping us; heroic, yes, battling for her children, working, sacrificing." This concept, of the Mother becoming somehow sexless, has been challenged by some writers, including Nancy Friday. Yet many daughters would prefer the stalwart Liebling mother to the mod mother portrayed by Linda Brooks in her 1981 picture entitled *Mom at 55, me nearing 30* (page 28). "Mom" is flirting with the camera, offering herself as strong competition to her daughter. "Mom" is also a social butterfly: can we ignore the butterfly motif on plants and in petit point? The Linda Brooks picture, like so many others in this collection, can be read as a painful reminder of the perils inherent in the mother-daughter relationship. If "Mom" is always outcharming the daughter, constantly demonstrating how youthful she is, how much prettier she is, then she cannot be used as a role model. She becomes the devouring

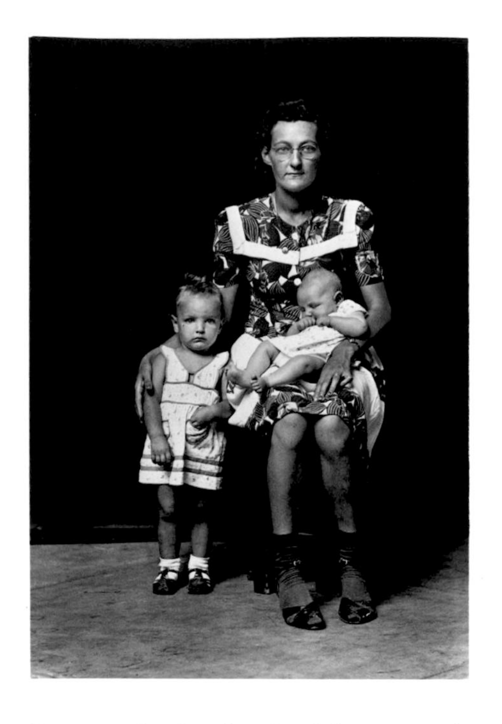

(left) Michael Disfarmer, *Donna Faye Latacher and children,* undated

Michael Disfarmer; *Selma McCarty and Wincie,* 1944

Gorgon of a daughter's nightmares, the enemy who is always victorious, the diminisher of one's hopes, the inflicter of a permanently damaged self-respect.

How different is the reverse of the problem! In Joel Meyerowitz's splendid portrait of an upper-class mother and daughter seen in the pastel light of an early evening on Cape Cod, it is the daughter rather than the mother who flirts with the photographer (page 11). What are the clues to their social class? It seems that it is only lower middle-class women and the jet set who bother to create fantastic bouffant coiffures. Intellectuals and old money disguise themselves, or simply do not care. Then there are the rings on the mother's hands, the casual but stunning white trousers, the Wasp features, her stance. The daughter, who seems accustomed to being admired, is carefully making a display of herself as elegant as any Botticelli. This is no candid shot taken with a 35mm autofocus, but a deliberate posing for an 8 x 10 view camera. These are two women who are presenting their personae more as individuals than as connected in a relationship: the mother's arm crosses behind the girl, but by no stretch of the imagination can this be considered an embrace.

Sexuality haunts all mother-daughter relationships. Daughters may be wonderful creatures, but their chastity must be protected. They must be safely married off. In a surreal moment captured by Sage Sohier at a Utah picnic (page 38), the sexuality of the daughter is overwhelming. *Zaftig* in her tight shorts, her blonde hair askew as if it were mirroring an inner wish for freedom from constraint, she holds—yes!—an apple to her mouth. Was there a satanic serpent in that tree? Eve: the eternal seductress, the first sinner, our first mother (unless you count Mother Earth, Lilith, or Pandora), who was called "The Mother of all Living Things," has also been defended by skillful exegesis of the book of Genesis, where, curiously enough, there are two versions of her creation. In the first, God creates male and female as equals; in the second, Eve is notoriously Adam's Rib. Apple to mouth, Sohier's young woman is a pictorial equivalent to that Eve who ate the fruit out of primordial disobedience, and who thereupon brought the knowledge of sexuality into the world. In that act, too, she cast herself not only into demeaning bondage to the male, but was cursed with having to bring forth children in dire pain. This story of the origins of male/female relationships and of the presence of evil has powerfully influenced attitudes even today, and still surfaces in the iconography of women.

Sexuality does indeed haunt all family relationships, even when the sexuality of the parents has been channeled into reproduction, as we see in Sally Mann's *Jenny and Her Mother* (page 25). An adolescent daughter clings to her pregnant mother, who leans against one of the symbols of womanhood: the washing machine. Why such sorrow in the daughter's face? Thirteen, perhaps, she is all too conscious of the genesis of that baby-to-be. Needy still, as are all adolescents who perch so precariously between the safety of childhood and the perils of young adulthood, she intertwines her body with her mother's. It is as if the daughter is drowning and is hanging on for dear life, while the mother, accustomed to the emotional demands of the child, strives to keep her head up, as if she is afraid of being sucked down by the child. Any excessively symbiotic relationship between mothers and daughters is threatening to each, for each needs to preserve her own identity and separateness. The Mann portrait is a marvelous display of the passion of the quintessential female-to-female arc.

Mothers are our first teachers, and it is they who decide what shall be our first lessons. Todd Merrill's *Susan and Blaze* (page 87) brings us into a contemporary middle-class kitchen. A young, attractive mother, her waist bound with a short red apron, is introducing her gap-toothed daughter to the single most important function of women in any family: the provi-

Dorothea Lange,
Migrant Mother,
Nipomo County,
California, 1936

sion of food. The girl is happy, excited; the mother pulls her daughter's head close, in a gesture that could smother the child. Together they make pizza. The girl is learning how to prepare the appetizing food that Daddy will enjoy with them when he comes home. It can be noted that most food advertising on television and in the popular magazines tends to equate food with love, with mother love in particular, as well as with sensuous pleasure. Mother, of course, cannot escape being equated with food: she *is* food, was our first food. As every marketing consultant for food companies knows, Mother is not only the vessel in which the fetus grows but also the vessel out of which food miraculously appears for the infant.

Some mothers do not necessarily concentrate on teaching their daughters how to cook or even how to become motherly. Some, as in Susan Copen-Oken's devastating photograph called *Coach* (page 73) are preparing their daughters to perform, to compete, perhaps for Miss America, perhaps for the musical comedy stage. Only someone who has ambitions to be on the stage herself would be able to teach her daughter how to mimic the desired gestures and poses. It seems natural enough for a parent to want to create a child in his or her own image. In *Mommie Dearest*, the scalding autobiography of Joan Crawford's daughter, Christina, there is a photograph of them together when Christina was perhaps four years old. They are dressed exactly alike, in an awfully folksy item of the fifties, dirndle jumpers over puffed-sleeve white blouses. Mother Joan, then a famous actress and glamour girl, obviously outshines the grinning tot. There is a hypothesis that women who create these performances of dressing alike were probably extremely frustrated in their relationships with their own mothers, and seek a close relationship with their daughters as compensation, also believing they will gain in youthfulness by such tactics as dressing their daughters in miniature versions of their own clothes.

Dressing alike may seem like fun to some daughters, but others want to keep their mothers from imitating their activities. Such a pair can be studied in Margaret Randall's *Mother and Daughter Riding* (page 79). It is impossible to guess whether it was the daughter or the mother who initiated this biking form of togetherness, yet the clue of the daughter's expression indicates that she would just as soon not be compared with her mother or to share the fun with the older woman, who seems to be having such a grand time. Each generation needs to find itself through generationally exclusive activities, language, clothing, and hair styles. To be too closely identified with the mother is viewed as not being "cool." Besides, the point of generationally exclusive behavior is to make separation from the parent possible—being a pal is not the same as being a Mom.

Sometimes a photographer has the good fortune to encapsulate the entire history of a relationship in a single picture. Carla Weber's acerbic *Donna and Diane* (page 50) reveals much more than the typical middle-class delight in portraiture. The pictures of daughter Diane, growing up blonde and once quite pretty, guard both sides of a mirror in which today's realities can be seen. The mother is forceful, with the beady eyes and jutting-forward gesture of her chin that makes her resemble a bird of prey. The daughter seems repressed, subdued, even close to depression. It is possible to read a failed life in the daughter's expression, and a wariness like a caged animal. These two are overwhelmed by the objects in the mother's house. One longs to say to these women: at least hold hands, look at each other, relate to each other, let the mother give the daughter the assurance that she need not live up to those banal portraits behind her.

After the airless claustrophobia of the Carla Weber portrait, turning to Milton Rogovin's dual image of an Appalachian woman and her daughter is a relief (page 84). Here is a woman who has had the courage to enter what has been traditional-

Eve Arnold,
Sailor at Home on Leave with his Family, 1950

ly a man's job: mining. Recent history tells us that she will have undergone a painful initiation, including rejection by her male coworkers, then perhaps a grudging respect. This woman is earning far more at a dangerous, grimy job than she might as a clerk in a local department store. Despite having entered a masculine province, when she washes up and sits in her living room embracing her reluctant daughter, she clearly retains her femininity. She even stresses the conventionality of her femaleness, if you accept as a clue the customary southern-belle doll atop the radio. Visible pride is here, and a comfortable affection for the child, but unfortunately the photographer can tell us little about the child except her recalcitrance at being photographed. We can guess that she may be one of those children who hates being hugged by her parents in front of strangers.

Paul Fusco's *Cocopah Indian Family* (page 61) is a revelation of another kind. The pain of poverty seems to have been overcome by a strongly loving connection. In what was apparently a grab shot, Fusco recorded a lively truth: the boys will be independent of their mother's affections far sooner than the girl will be. How very easy they are with each other, this mother and her pretty daughter; the boys, meanwhile, are roughhousing, learning how to be men as they might understand the meaning of that role. When a photographer has been completely accepted by a family, when they become oblivious to his or her presence, forgetting the clicking of the shutter, wonderfully truthful and intimate insights can be gleaned.

One of the great masters of this kind of insight, Garry Winogrand, gives us a glimpse of an eternal verity (page 107) with his picture of a young mother with one child in a shoulder carrier and her daughter embracing her hand with an expression of total adoration. Watching her step on the often potholed pavements of Manhattan's Fifth Avenue, the mother seems somewhat abstracted, yet she is not aloof. She holds the girl's hand firmly, protectively. It is easy to imagine this mother and this daughter sharing a story over milk and cookies, or reading aloud together from a book before the girl goes to sleep. So many photographs of mothers and daughters, from the time of Gertrude Käsebier to the present, display them in this storybook relationship. Alice Walker, author of *The Color Purple*, wrote about her own mother's stories as the most precious cargo of her soul. It was not only that she would retell them, but that she believed her mother was a writer *manqué*. What was more important, claimed Walker, was the fact that she absorbed the cadences of her mother's voice until they became part of her own art. Even though the Winogrand photograph does not show a mother reading to a child, something about the daughters' almost convulsive gesture of love bespeaks a close, articulate friendship between the two.

Perhaps the most joyous photograph in the entire collection is Bea Nettles' *Rachel and the Bananaquit* (front cover). Bea has undoubtedly been telling her merry daughter about a fantastic bird, a story that Rachel will remember and can recall as long as this picture survives. Here is the mother-daughter combination before the pressures of growing up intervene, before competition emerges, before sexuality threatens, before illness or separation or divorce or social disaster can overpower this captivating *joie de vivre*. It is perhaps the single picture in the collection that is a genuine Adoration, with the daughter even raising her chubby little hand as if in benediction. It is an incredibly rich image despite its pictorial directness and simplicity.

Could the Nettles' picture have had the same effect in black and white? No. An immediate loss would be the warm interaction between flesh tones, fantasy creature, and tropical landscape. Yet many photographers contend that nothing but black and white can express concepts. Robert Frank, for one, once insisted that black and white were the only possible colors for

Jerome Liebling, *Wedding*, Brooklyn, New York, 1947

photographs. He wrote that they symbolized the antipodes of hope and despair to which humanity was forever doomed. His was a tragic vision, yet not all photographers choose to see the world in such radically opposed values. Even Frank admitted, "There is one thing the photograph must contain, the humanity of the moment." If several of the photographs in this collection hint at despair, many of them, like the Nettles picture, do indeed reverberate with the humanity of the moment, humanity here meaning not simply people, but their humaneness, their tenderness, their loving, their fragilities, their strengths.

Since our humanity can be expressed in both verbal and visual languages, it can be said that *Mother* is a verb, while *Daughter* can be nothing but a noun. *To mother* is to nurture, to cherish, to nurse, to care for, to encourage, to provide sustenance for, to educate: those are the good attributes of our verb. The less admirable attributes are still verbs: to dominate, to squelch, to crush, to compete, to smother—yet these are not ordinarily considered the exclusive behaviors of the role of Mother, but merely the unsavory behaviors of human beings, male or female. *To daughter* does not exist as a verb; one cannot "daughter" someone else, whereas one can "mother" even if you are male.

This difference in the syntactical definitions of their roles accounts for the absence of some types of interactions from this collection. We do not, for example, see many older daughters caring for their invalid mothers. We do not see—and how to represent it?—mature women taking on the reversed roles of mothering the mother in her old age. We do, however, see glimpses of the generations of great grandmother, grandmother, mother, and daughter, with the daughter sometimes bearing in her arms the succeeding generation. This chain of being, female to female, has been the inspiration for much contemporary literature and poetry, and, to judge by these photographs, has become the focus of attention for many visual artists as well. The struggle is to convey through concrete physical appearance the ineffable, the indescribable subtleties of a primary female relationship.

That relationship between mothers and daughters has undergone many changes. While every mother in the past was perforce also a daughter, today's options ensure that not every daughter may choose to be a mother. It was perhaps inevitable that sophisticated technology, birth control methods that are relatively safe, a complete alteration in the nature of work and other socioeconomic factors of the postindustrial world, would impinge on the structure of the family. It was inevitable that the division of labor eminently suited to the cave-hunter stage of society—with women safely indoors bearing and rearing children, and men outdoors hunting the woolly mammoth for food—would alter not only with the advent of agriculture and the domestication of animals but also with the long history of a shift from production to service. The complexities of cities, the long commutes to work, the recent struggle for the father to share the responsibilities of parenthood, the ever-rising rate of divorce and the subsequent abandonment of women to poverty-stricken single parenthood—all these factors will impact on what might otherwise seem to be the most natural and fundamental of all relationships, that of mother to child. Even the privilege of educating the daughters may be shifted elsewhere.

Photographers will need to be able to communicate these subtleties without reviving traditional stereotypes. So much is invisible in any relationship that it requires great feats of the imagination to conjure up significant meanings that we can share. We are still learning about the varieties of mother-daughter interactions, just as we are beginning to learn about many other aspects of what it means to be human. These pictures demonstrate that there is no single mother-daughter relationship, but almost as many differences as there are individuals.

THE ENVELOPE

It is true, Martin Heidegger, as you have written,
I fear to cease, even knowing that at the hour
of my death my daughters will absorb me, even
knowing they will carry me about forever
inside them, an arrested fetus, even as I carry
the ghost of my mother under my navel, a nervy
little androgynous person, a miracle
folded in lotus position.

Like those old pear-shaped Russian dolls that open
at the middle to reveal another and another, down
to the peasized, irreducible minim,
may we carry our mothers forth in our bellies.
May we, borne onward by our daughters, ride
in the Envelope of Almost-Infinity,
that chain letter good for the next twenty-five
thousand days of their lives.

MAXINE KUMIN

107 Garry Winogrand, *Untitled*, New York, New York, 1967

NOTES ON THE EXHIBITION

In January 1986, Aperture began work on the Mother-Daughter project, which we hoped would describe the changing nature of women in American society and further examine how those changes had altered or affirmed the traditional familial bond.

More than 600 photographers worldwide, both well-known and those less prominent, were asked to submit their personal choices of images of American mothers and daughters for consideration in the show. Many of these artists in turn told others of the project, or wrote and telephoned with their suggestions for important material. The resulting preliminary choice of some 3000 photographs by contemporary photographers was exhilarating in its breadth and depth. To make the final selection of images was a difficult and lengthy decision-making process.

We discovered many photographers whose work on the mother-daughter relationship had been ongoing for years. We uncovered artists who had created collages, albums, books, series images, and other treatments in their search for expression on this theme. Photographers wrote of their excitement that an exhibition was finally being planned on one of the central concerns in their work. They wrote of themselves, and their personal lives—to an unprecedented degree—and we asked permission to include these comments with the images.

In choosing the final photographs for the exhibition we considered both the visual merit of the images, and the breadth and diversity of the subject: the classes, races, styles, and ethnic identities that mother-daughter relations encompass in America. We looked for images that had something to say, that shouted some feeling or whispered emotion. We looked for the small details that might fall between the cracks. The exhibition was envisioned as a tribute to American women: in this potpourri of images, text, commentary by the artists, and other materials, we hope that we have selected work that reflects the scope and intensity we encountered, and that will provide a point of departure for future visual explorations of mothers and daughters.

DIANE LYON AND NAN RICHARDSON

Because of space restrictions, the work of the following photographers who are represented in the *Mothers & Daughters* exhibition was not able to be reproduced in the catalog:

Joan Albert, William Albert Allard, Tina Barney, John Carrino, John Cavanagh, Judy Dater, Jed Devine, Patricia Evans, Leonard Freed, Mary Frey, Patricia Galagan, Penny Gentieu, Jim Goldberg, Nan Goldin, Susan Kae Grant, William Hubbell, Tamara Kaida, Bud Lee, Phyllis Leideker, Helen Levitt, John Lueders-Booth, Jeff Mermelstein, Joan Moss, Phil Norwich, Wendy Olson, Melissa Pinney, Sarah Putnam, Jack Radcliffe

ACKNOWLEDGMENTS

We are indebted to Carole Kismaric for her invaluable advice in picture research; to Frances Fralin for her warmly appreciated recommendations; to Larry Frascella, George Slade and David Lee for their editorial assistance at Aperture; to Mary Virginia Swanson for her timely assistance in contacting photographers; to Peter A. Andersen and Elizabeth J. McCoy for their guidance in systemizing the deluge of photographs; to Sally Fryberger and Renee Hughes for all their help. Thanks also to Jacqueline Leclerc at the Library of Congress, to Margaret Sidlovsky at Magnum Photos, and to Terry Barbero at Archive Pictures, who negotiated red tape with efficiency and aplomb.

PHOTO CREDITS p. 5, courtesy Magnum Photos Inc., New York: p. 33, collection of Sallie Norquist, Ph.D. and John Filak, Ph.D.; pps. 46, 52, 53, courtesy Archive Pictures, Inc., New York; p. 54, courtesy Pace/MacGill Gallery, New York; pps. 59, 60, 61, courtesy Magnum Photos Inc., New York; p. 62, courtesy Woodfin-Camp, New York; p. 68, courtesy Archive Pictures, Inc., New York; p. 73, courtesy Magnum Photos Inc., New York; p. 75, courtesy Archive Pictures, Inc., New York; p. 91, courtesy Pace/MacGill Gallery, New York; p. 99, left: courtesy of The Library of Congress; p. 99, right: courtesy of the International Museum of Photography at the George Eastman House, Rochester; p. 100, left and right, courtesy of The Library of Congress; p. 101, left: courtesy of Naomi and Walter Rosenblum; p. 101, right: courtesy of Mrs. Donna Van der Zee; p. 102, left and right, courtesy of the Staley-Wise Gallery, New York; p. 103, left: courtesy of the Dorothea Lange Collection at the Oakland Museum of Art; p. 104, right: courtesy of Magnum Photos Inc., New York; p. 105, courtesy Jerome Leibling.

CONTRIBUTORS

Tillie Olsen is the author of *Tell Me a Riddle, Yonnondio, Mother to Daughter: Daughter to Mother,* and *Silences.* She has received numerous awards and honors, including the O. Henry Award for the best short story in 1961, a John Simon Guggenheim Memorial Fellowship, and honorary degrees from several colleges and universities, including the University of Nebraska. Tillie Olsen lives in California with her husband, near their four daughters and seven grandchildren.

Julie Olsen Edwards, the second of Tillie Olsen's four daughters, teaches Child Development and Women's Studies at Cabrillo Community College, Santa Cruz, and does public advocacy work for young children and their families. She has lectured widely on the subject of motherhood, and has just published her first work, "Mother Oath." She lives with her husband and two children in California.

Estelle Jussim is author of the award-winning titles *Landscape as Photograph* (Yale, 1985), *Slave to Beauty* (Godine, 1981), *Frederic Remington, the Camera and the Old West* (Amon Carter Museum/Tandy Lectures in American Civilization, 1983) and *Visual Communication and the Graphic Arts* (Bowker, 1974), as well as monographs on Jerome Leibling, Barbara Crane, and others. Dr. Jussim is a professor on the graduate faculty of Simmons College, Boston, and a highly respected commentator on aesthetics, photography and film, and popular imagery. Recipient of a John Simon Guggenheim Memorial Fellowship in 1983 for work on Alvin Langdon Coburn, she is currently engaged in research on documentary film and photography.

In celebration of Aperture's 35th anniversary

A SPECIAL OFFER TO SUBSCRIBERS-

For the collector or the connoisseur of fine photography, Aperture offers a series of special editions. Elegantly bound and slipcased, with foil stamping, each book is accompanied by an original signed print or hand-pulled gravure of the photographer's most representative work. The print is provided by the photographer or under terms dictated by the photographer's estate or by expert printmakers working in cooperation with art historians and scholars. Each print is matted on archival board and presented in a special folio. Aperture special collector's editions provide an unmatched opportunity to possess the finest product of the bookmaker's craft, and to enjoy the exquisite quality of a great photographic print. The rising prices of photographs and art photography books suggests that the purchaser of a special edition may reasonably expect it to increase in value in the years ahead. In celebration of Aperture's 35th anniversary, subscribers may receive a 50 percent discount when ordering before 12/31/87. New York State residents add applicable sales tax. To place an order, call or send your payment to Aperture, 20 East 23rd Street, New York, NY 10010, 212/505-5555. Limited quantities are available. The collector's editions are available for viewing by appointment with the Burden Gallery at Aperture.

Robert Adams:
From the Missouri West
Edition limited to 100 numbered copies, signed by Robert Adams. An original silver print, *Bulldozed Slash, Tillamook County, Oregon,* 1977 (image 9 x 7 in.), accompanies the edition. Each print is signed by the photographer.

$400.00 ISBN 0-89381-063-0

Shirley C. Burden:
Presence
Edition limited to 100 numbered copies, signed by Shirley C. Burden. An original silver print, *Ellis Island,* 1956 (image $9\frac{1}{4}$ x 12 in.), accompanies the edition. Each print is signed by the photographer.

$250.00 ISBN 0-89381-076-2

William Christenberry:
Southern Photographs
Edition limited to 100 numbered copies, signed by William Christenberry. An original Ektacolor print, *House and Car, near Akron, Alabama,* 1981 (image 8 x 10 in.), accompanies the edition. Each print is signed by the photographer.

$300.00 ISBN 0-89381-130-0

Larry Fink:
Social Graces
Edition limited to 250 numbered copies, signed by Larry Fink. An original silver print, *Tavern on the Green, New York City,* 1976 (image 9 x $10\frac{1}{2}$ in.), accompanies the edition. Each print is signed by the photographer.

$250.00 ISBN 0-89381-159-9

A portfolio of 82 silver prints, produced by Larry Fink and limited to 30 examples and 7 artist's proofs, is available. Write Aperture for details.

Randal Levenson:
In Search of the Monkey Girl
Edition limited to 100 numbered copies, signed by Randal Levenson. An original silver print, *Barbara Bennett, "World's Smallest Mother," and Ed Bennett, Columbus, Ohio,* 1976 (image $18\frac{1}{2}$ x $14\frac{1}{2}$ in.), accompanies the edition. Each print is signed by the photographer.

$300.00 ISBN 0-89381-097-5

A collector's edition of five original prints is also available.

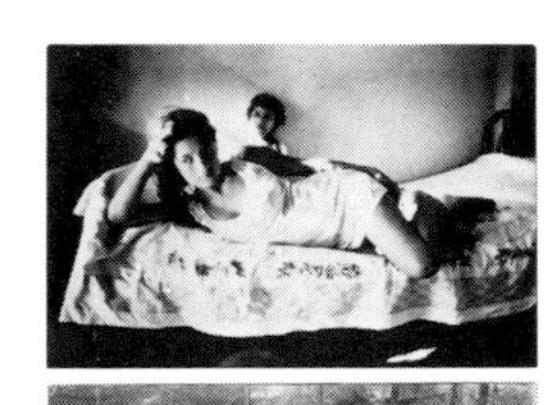

Danny Lyon:
Pictures from the New World
Edition limited to 400 numbered copies, signed by Danny Lyon. An original silver print, *Gloria and Rosario, Santa Marta,* 1972, accompanies numbers 1-200; *Ellis Unit,* 1968, accompanies numbers 201-400 (both images $7\frac{3}{4}$ x $11\frac{1}{3}$ in.). Each print is signed by the photographer.

$400.00 ISBN 0-89381-082-7

Lisette Model:
An Aperture Monograph
Edition limited to 300 numbered copies, signed by Lisette Model. An original silver print, *Sailor and Girl,* c. 1940 (image 16 x 20 in.), accompanies the edition. Each print is signed by the photographer.

$300.00 ISBN 0-89381-052-5

Timothy H. O'Sullivan:
American Frontiers, 1867-1874
Edition limited to 200 numbered copies, accompanied by an albumen print made by means of the original negative and authorized by the National Archives and Records Service, Washington, D.C. *Shoshone Falls, Snake River, Idaho,* 1874, accompanies

50% DISCOUNT ON COLLECTOR'S EDITIONS

numbers 1-100; *Cooley's Park, Sierra Blanca Range, Arizona*, 1873, accompanies numbers 101-200 (both images 10 x 12¼ in.).

$250.00 ISBN 0-89381-094-0

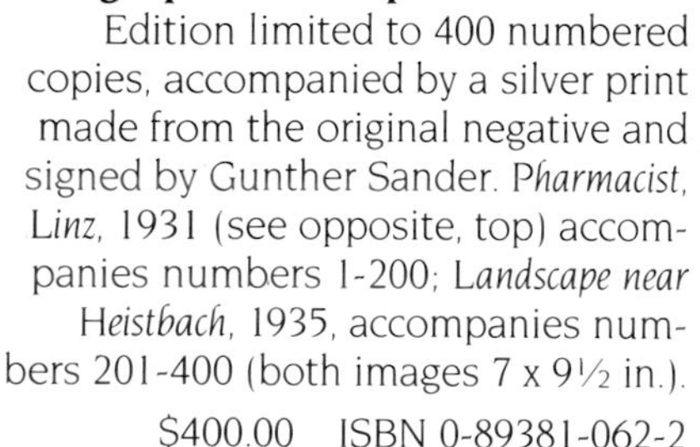

August Sander:
Photographs of an Epoch 1904-1959
Edition limited to 400 numbered copies, accompanied by a silver print made from the original negative and signed by Gunther Sander. *Pharmacist, Linz*, 1931 (see opposite, top) accompanies numbers 1-200; *Landscape near Heistbach*, 1935, accompanies numbers 201-400 (both images 7 x 9½ in.).

$400.00 ISBN 0-89381-062-2

Stephen Shore:
Uncommon Places
Edition limited to 100 numbered copies. An original Ektacolor print, *Merced River, Yosemite National Park, California*, August 13, 1979 (image 7¾ x 9¾ in.), accompanies the edition. Each print is signed by the photographer.

$250.00 ISBN 0-89381-104-1

W. Eugene Smith:
Master of the Photographic Essay
Edition limited to 1,000 numbered copies. A facsimile hand-pulled gravure, *The Spinner*, 1950 (image 9 x 12 in.), accompanies the edition. Each gravure is approved by John Morris on behalf of the Estate of W. Eugene Smith and signed on the cover sheet.

$175.00 ISBN 0-89381-072-X

Brett Weston:
Photographs from Five Decades
Edition limited to 400 copies, signed and numbered by Brett Weston. An original silver print, *Reeds, Oregon*, 1975 (image 11 x 14 in.), accompanies this edition. Each print has been signed by the photographer.

$800.00 ISBN 0-89381-068-1

Paul Strand:
Sixty Years of Photographs
Edition limited to 350 numbered copies. An original hand-pulled gravure, *Old Fisherman, Gaspé*, 1936 (image 7 x 5½ in.), accompanies the edition.

$150.00 ISBN 0-89381-011-8

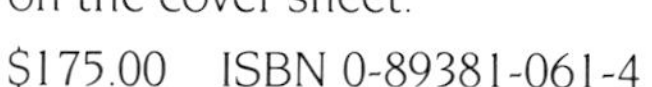

Paul Strand:
Time in New England
Edition limited to 450 numbered copies. An original hand-pulled gravure, *Iris*, 1928 (image 9½ x 7½ in.), accompanies the edition. Each gravure is approved by Hazel Strand on behalf of the Estate of Paul Strand and signed on the cover sheet.

$175.00 ISBN 0-89381-061-4

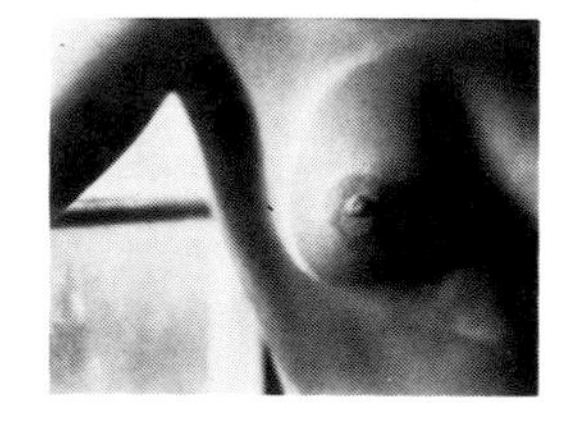

Edward Weston:
Nudes
Edition limited to 350 numbered copies, signed by Charis Wilson. A platinum print, *Nude*, 1920 (image 7¼ x 9¼ in.), made from the original negative under Cole Weston's supervision, accompanies the edition. Each print is signed by Cole Weston.

$425.00 ISBN 0-89381-025-8

Edward Weston:
California and the West
Edition limited to 350 numbered copies, signed by the author Charis Wilson. A silver print, *Juniper, Lake Tenaya*, 1937 (image 9½ x 7½ in.), made from the original negative by Cole Weston, accompanies this edition. Each print is signed by Cole Weston.

$400.00 ISBN 0-89381-037-1

Edward Weston:
His Life and Photographs
Edition limited to 350 numbered copies, signed by Cole Weston. A silver print, made from the original negative by Cole Weston, *China Cove, Point Lobos*, 1940 (image 7½ x 9½ in.), accompanies the edition. Each print is signed and numbered by Cole Weston.

$450.00 ISBN 0-89381-045-2

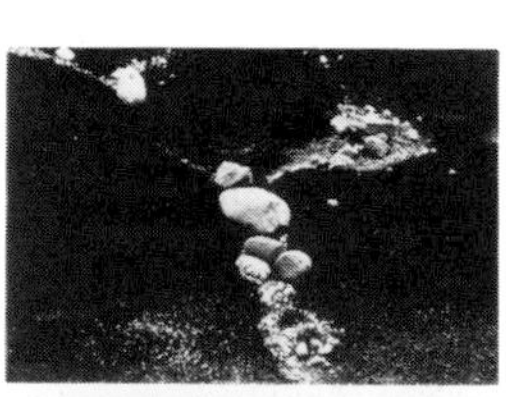

Minor White:
Mirrors/Messages/Manifestations
Edition limited to 50 numbered, unsigned copies. An original print, made by the photographer, accompanies the edition. *Stone Steps, Notom, Utah*, 1967 (image 7½ x 10¾ in.), accompanies numbers 1-25; *Rings and Roses, Ponce, Puerto Rico*, 1973 (image 9¼ x 11¾ in.), accompanies numbers 26-50. No original negative exists for *Rings and Roses*.

$900.00 ISBN 0-89381-105-X